Controversial Issues in Anglo-Irish Relations, 1910–1921

Controversial Issues in Anglo-Irish Relations, 1910–1921

Cornelius O'Leary & Patrick Maume

FOUR COURTS PRESS

Set in 10.5 on 13 point Times New Roman for
FOUR COURTS PRESS LTD
7 Malpas Street, Dublin 8, Ireland
email: info@four-courts-press.ie
http://www.four-courts-press.ie
and in North America for
FOUR COURTS PRESS
c/o ISBS, 920 N.E. 58th Avenue, Suite 300, Portland, OR 97213.

ISBN 1–85182–657–2

A catalogue record for this title
is available from the British Library.

Printed in Great Britain
by Antony Rowe Ltd, Chippenham, Wilts.

Contents

Preface

Why another book on Anglo-Irish relations between 1910 and 1921, on the Home Rule crisis and the Treaty negotiations? Our purpose is not to recapitulate the arguments of the standard surveys of modern Irish history by F.S.L. Lyons, Roy Foster and J.J. Lee, or even of specialist works like Nicholas Mansergh's *Unresolved Question*, Patrick Maume's *Long Gestation*, or Michael Hopkinson's *Irish War of Independence*, but rather to examine questions which in our opinion have not been adequately answered by previous scholars – for example, whether serious danger existed of a civil war in Ireland in the summer of 1914; whether there was a real agreement between John Redmond and Edward Carson in the summer of 1916; what were the implications of the extraordinary elections of 1918 and 1920; how far did Ulster Unionist opposition affect the Treaty negotiations in 1921?

Students of this subject have had in recent years access to sources not available in the early decades. The relevant British cabinet papers were only made available to scholars in the 1960s, and even yet Dominion Office papers are being released. Significant diaries and other private papers have recently become accessible. The publication of Tom Jones's *Whitehall Diary* in 1971 provided an insider's account of British cabinet proceedings in 1921; the perspective of a Dublin Castle official in the dying days of the old regime has been made available with the 1999 publication of the Mark Sturgis diaries, while Eamon Phoenix's *Northern Nationalism* (1994) was the first scholarly account of a major factor in the negotiations of the period.

It is not our purpose to give in this preface a précis of the entire argument. We leave it to the readers to decide whether our account of the centrality of the Ulster question is satisfactory.

We wish to thank the following for access to copyright materials, the Directors of the Public Record Office at Kew, the British Library, the National Archive of Ireland, Dublin, and the Plunkett Foundation, Long Hanborough, Oxfordshire; the Clerk of the Records, House of Lords Record Office, Bodley's Librarian, the Librarian of Nuffield College, Oxford; the Librarian of Birmingham University Library, the Librarian of Trinity College Dublin, and the Head of the Archives Department, University College Dublin. Asquith to George V, 27 June 1916 (CAB 41/37/24) is published from photographic copies in the National Archives of original letters preserved in the Royal Archives and made available by gracious permission of Her Majesty the Queen.

We wish to thank our colleagues who have commented on parts of the manuscripts – Paul Bew, George Boyce, Tom Garvin, Alvin Jackson. For any errors of fact or judgment we are solely responsible.

C. O'L & P.M.

September 2004.

1

The home rule crisis, 1910–14

At the beginning of 1909 there was nothing to suggest that the Irish problem, 'that damnable question', was about to enter its most acute phase, and that every year between 1910 and 1922 (1915 excepted) would see at least one British government 'initiative'. The Irish Council bill of 1907 which would have provided a measure of devolution for some major services had been rejected by the Irish Party, who decided against a gradualist approach to home rule, but they had welcomed the Irish Universities bill of 1908, which established the federal National University of Ireland, with three constituent colleges and the unitary Queen's University of Belfast. The Liberal government of Asquith, who had assumed the premiership on the resignation of Campbell-Bannerman in April 1908, at first seemed to adhere to the same policy of 'step-by-step' advances towards home rule.

Several historians tend to exaggerate the closeness of the relationship between the Liberals and Irish Nationalists at this time. To use Mansergh's felicitous phrase, it was an 'entente' but not 'cordiale'.[1] It became distinctly less than cordial with the introduction of Lloyd George's 'People's Budget' in 1909; the Irish members approved (generally) of the new welfare programme, but were deeply disappointed at the liquor tax, which seriously alienated one of the strongest interests backing the party. After futile efforts to secure modification to the bill, the Irish members voted against the second reading, but, aware of the anomaly of lining up with the Unionists, they abstained on the third reading in the Commons, which was passed by 379 votes to 149.

After the rejection of the budget by the 'last-ditch' majority in the House of Lords (350 votes to 75), a dissolution immediately followed (December 1909). The Irish Party were deeply conscious of discontent at home about the slow progress of 'the cause', and even before the dissolution Redmond wrote a strong letter to John Morley (the most prominent surviving old-style Gladstonian home ruler in the cabinet), demanding a public commitment to home rule, if the Irish vote were not to be thrown against the Liberals in the coming campaign. On 10 December, in a speech at the Albert Hall, Asquith stated categorically that the Irish problem could only be solved 'by a policy which, while (explicitly) safeguarding the supremacy and indefeasible authority of the Imperial parliament,

---

1 Nicholas Mansergh, *The Unresolved Question: The Anglo-Irish Settlement and Its Undoing* (New Haven and London, 1991), p. 121.

will set up in Ireland a system of full self-government in regard to purely Irish affairs'.[2]

Fortified by this pledge, the Redmondites turned their attention to their own internal squabbles and won 70 of the 103 Irish seats, while 11 independent Nationalists (mainly grouped around the dissidents, O'Brien and Healy), one Liberal home ruler and 21 Unionists made up the rest. But the overall results, Liberals 275 and Unionists 273, gave the Irish (with the 40 Labour representatives) the balance of power.

This advantage they were not slow to exploit. After the election the first priority for the government was to pass a budget in place of the defeated 'People's Budget'. Early in February 1910 Asquith was informed that the Irish members would not allow the budget to pass unless a pledge were given to introduce a measure limiting the veto power of the Lords during that session.[3] The government replied by prevaricating; although some members thought that the most dignified course would be to resign and force another dissolution. The next question was whether the Irish would support the budget if it were introduced in the interval between the veto resolutions being considered by the two houses. Redmond held firm to the slogan 'No veto, no budget'. Both sides were well aware that a veto bill would have no chance, unless accompanied by a pledge from the king to create enough peers to overcome the enormous Conservative majority in the Lords. At last, in spite of public protestations that he would not concede the Irish demands, Asquith informed the king that if, as expected, the Lords rejected the forthcoming Veto bill, he would ask for a dissolution, and requested that the king guarantee in the event of a Liberal victory, to create enough peers to carry the bill. Edward VII, in almost his last public act, agreed. When Asquith conveyed the royal assurance to the Commons, the Irish made no further difficulties about the budget, which passed into law by 27 April. (Three resolutions embodying the principle of the future Parliament bill, the abolition of the Lords' veto, had passed the Commons on 14 April.)

So when parliament adjourned for the Easter recess in 1910, the stage had been set for the greatest constitutional crisis in British history since the end of the seventeenth century, leading to the emasculation of the unrepresentative house of peers. For this development Irish nationalists could claim most of the credit. It was Redmond and his colleagues who successfully pressed home the advantage won in the January election. Patrick O'Farrell's generalization that the Irish party was 'entirely at the mercy of the British political situation',[3] however true of earlier periods, is emphatically not applicable to the year 1910 to which specifically he refers. Foster also harps excessively on Redmond's difficulties.[4] The

2 J.A. Spender and Cyril Asquith, *Life of Lord Oxford and Asquith* (London, 1932), pp 268–9.   3 Patrick O'Farrell, *Ireland's English Question* (New York, 1972), p. 260.   4 R.F. Foster, *Modern Ireland, 1600–1972* (London, 1989), p. 462.

most accurate assessment seems to be that of Mansergh in his posthumous *The Unresolved Question.* 'By the end of 1910 Redmond had almost every cause for satisfaction.'[5]

But this is to anticipate. By June 1910, when the Parliament bill was introduced in the Lords, Asquith secured from the new king, George V, who had succeeded to the throne in the previous month, a pledge that if the Lords defeated the bill and (he) Asquith asked for a dissolution and the Liberals were returned to power, the king would create sufficient peers to secure the passing of the bill. In return, the king urged the calling of a conference between the party leaders to explore the possibility of agreement on constitutional questions and so avoid another election. The discussions were confined to the leaders of both main parties, the Labour and Irish members being excluded, although the Irish chief secretary, Birrell, was one of the Liberal delegation. The conference continued until November and then petered out. The Conservatives, having allowed themselves to be distracted by a quixotic proposal by Lloyd George for a coalition government,[6] found the Liberals adamant, with home rule obviously in mind, in refusing to countenance to a new category of 'constitutional' legislation subject to referendum. No agreement was reached, largely because the Conservative leaders, Balfour and Lansdowne, resolutely refused to accept either a settlement of the Commons versus Lords question, or of the Irish question in any way which in their opinion would imperil the unity either of the party or the empire.[7]

The Parliament bill, removing altogether the power of the Lords to delay financial measures and on general legislation prescribing a delaying power of two years, having passed the Commons, supported by all parties except the Tories, went to the Lords in June 1910. However, the peers used another type of delaying power, postponing consideration of the bill until they discussed real or pretended schemes for second chamber reform. To this Asquith responded by calling an election for December 1910 – the first occasion in which two general elections were held in the same year.

Although the Liberals and Irish home rulers fought the election on the issue of Lords versus people, with less emphasis on home rule, the Conservatives were preoccupied with an internal policy problem on which the party had agonized for nearly a decade, whether to solve the vexed question of tariff reform by a referendum. This issue split the leadership down the middle with Austen Chamberlain vehemently against, Bonar Law and Lansdowne for, and Balfour waver-

---

5 Mansergh, *The Unresolved Question*, p. 39.   **6** F.S.L. Lyons, *Ireland since the Famine* (London, 1968), p. 265. Catherine Shannon, *Arthur J. Balfour and Ireland, 1878–1922* (Washington DC, 1988), p. 153, points out that Balfour opposed a federal solution favoured at the time by Austen Chamberlain and J.L. Garvin of the *Observer*, partly on the ground that it would naturally tend towards disintegration of the UK, partly because the Irish would regard it as a stepping stone to full independence.   **7** The whole episode is discussed in David Dutton, '*His Majesty's Loyal Opposition*': *The Unionist Party in Opposition, 1905–1915* (Liverpool, 1992), pp 87–94.

ing. Eventually Balfour came down in favour of a referendum before any taxes on food would be imposed, but he wished to fight the election on the issue of the Union as a popular one around which the party could unite.

The outcome was a virtual repeat of the January result. The Liberals and Unionists tied with 272 seats each, Labour won 42, Redmond's majority Irish faction increased to 73; the independent Nationalists, while losing votes, won ten seats and one solitary Ulster Liberal made up the total. But while the Liberals and Irish were united internally on policy – the twin pillars of the Parliament bill and home rule – the Tories were in a state of disarray, well summed up by Austen Chamberlain:

> We had declared for the reform of the House of Lords, but there was no agreement as to the extent and character of the change to be made. We had adopted the Referendum, but even among those who welcomed it there was no agreement as to the circumstances in which it was to be applied. The Budget dispute was ended and the Budget which had raised such fears had become law, but we were now to face the battle over the Constitutional issue which the rejection of the Finance bill by the House of Lords had brought to a head.'[8]

On 21 February 1911 the Parliament bill (in identical form to the previous year's resolution) was introduced in the Commons. It reached the Lords on 23 May, and was met by the Tory leader, Lord Lansdowne's proposals for reform, including limiting the number of hereditary peers. But Lansdowne's proposals did not even appeal to his own backbenchers, and the next question was whether to let the bill through or stage another 'last-ditch' resistance. In spite of Asquith's clear statement in the previous year, Lansdowne and other Tories could not bring themselves to believe that the new king would engage in a mass creation of peers. The king was bombarded with contradictory advice from both parties and wryly said that whatever he did would be opposed by half the nation. Eventually, Lloyd George convinced even Lansdowne that the pledge to create peers was genuine.

At this point, Balfour and Lansdowne agreed, as a *faute de mieux*, to pass the bill. But for over a month they had to endure a last-ditch revolt, led by the 88-year-old Lord Halsbury, several times lord chancellor. Several leading Tory peers joined the diehards, supported by some Unionist frontbenchers in the commons, but they were balanced by influential former office-holders such as Curzon, ex-viceroy of India, and Lord St Aldwyn, who as Sir Michael Hicks-Beach had been chief secretary for Ireland and chancellor of the exchequer in the Salisbury cabinets. When the final division was taken in the Lords on 10 August 1911, 30 Tory peers voted with the government to ensure victory for the bill by 17 votes.

---

8 Austen Chamberlain, *Politics from Inside: An Epistolary Chronicle, 1906–1914* (London, 1936), p. 317.

In the immediate aftermath of the passing of the Parliament Act, Tory disaffection increased against Balfour for his ineffectual leadership generally, and especially over the Parliament bill. He was persuaded to resign as leader in November 1911. The two obvious candidates, Austen Chamberlain and Walter Long, had equally obvious defects: Chamberlain lacked a real will for power and pulled out of the contest at an early stage; Long, although he had held several high offices, including leadership of the Irish Unionists to whom he was closer than Bonar Law who was close to the Ulster Unionists, was, as Dutton says 'an undistinguished debater and not renowned for his intellectual gifts or quick wits'.[9]

In such a case, as in many others (Home, Lynch, Major), the choice fell on a third candidate, Andrew Bonar Law, with strong family connections to Ulster, born in Canada, a businessman in Glasgow, and MP only since 1900 and devoid of cabinet experience. More significant still was the election of Edward Carson, MP for Trinity College, Dublin, since 1892, as leader of the Irish Unionists in succession to Walter Long, who in December 1910 moved to an English constituency from Dublin County South. These two appointments, as will be seen, presaged much trouble for the home rule cause.

Nothing better displays the Hibernocentric tendency of Irish historians than the manner in which without exception they move from the passing of the Parliament bill in August 1911 to the preparation for and introduction of the Government of Ireland bill (the official title of the third Home Rule bill), without adverting at all to the fact that the cabinet had other serious issues on its mind. The Agadir incident in June 1911 poisoned Anglo-French relations and nearly led to war between France and Germany, while the miners were showing signs of industrial militancy. The committee assigned to draw up the bill used the bill of 1893 as a model, but the final draft was not ready until a few weeks before the bill was to be introduced (April 1912).

Meanwhile, however, the Ulster Unionists had clearly indicated the intensity with which their opposition would be conducted. Already there was talk of armed resistance. Carson, their new leader, whose cardinal political principle was the unity of the empire, had not played much of a part in the debates on the parliament bill. However, shortly after the Act became law, his chief lieutenant, Sir James Craig, organized a monster meeting of Ulster Unionists to introduce the new leader to his following. At that meeting, held on Craig's estate on 23 September 1911, some 50,000 men from all over the province heard the first of Carson's dire warnings of consequences, if the Liberal majority were to present home rule as a *fait accompli*.

---

9 Dutton, '*His Majesty's Loyal Opposition*', p. 165.

We must be prepared … the morning home rule passes, ourselves to become responsible for the government of the Protestant province of Ulster.'[10]

A few days later a meeting of delegates of Unionist and Orange institutions decided to appoint a small committee, including Carson and Craig, to frame a constitution for a provisional government for the nine counties of Ulster.

At that time the representation of the province at Westminster was evenly balanced, 17 Unionists to 15 Nationalists and one Liberal (dependent on Nationalist votes);[11] but the more important criterion for self-determination then used was the overall religious distribution in the population, since it was assumed that by and large Catholics would vote Nationalist and Protestants Unionist. The census figures for 1911, the most elaborate ever in Ireland, gave a breakdown for each county and for the province as a whole (see Table 1).

Table 1

Distribution of population by religion in Ulster, 1911[12]

County	Protestant (%)	Catholic (%)
Antrim	79.5	20.5
Down	68.4	31.6
Armagh	54.7	45.3
Londonderry	54.2	45.8
Tyrone	44.6	55.4
Fermanagh	43.8	56.2
Monaghan	25.3	74.7
Cavan	21.1	78.9
Donegal	18.5	81.5
Whole province	56.1	43.9

Irish historians have generally interpreted these figures in a one-sided manner, stressing the fact that Catholics were in a majority in Tyrone and Fermanagh, counties eventually included in Northern Ireland, while ignoring the fact that Protestants were in a clear majority over the province as a whole, and that that majority had been on the increase since 1886.

Lee is one of the few to advert to these facts, but glosses over them as follows:

10 A.T.Q. Stewart, *The Ulster Crisis* (London, 1967), p. 48.   11 In the elections of December 1910 the vagaries of the majority system returned Nationalists with very small majorities, and Unionists with large majorities.   12 *Census of Ireland*, 1911.

It was this sense of inalienable superiority that made Ulster Unionists impervious to the logic of numbers. The Ulster Unionist mind saw no incongruity in denying any nationalist right to rule the nine counties of 'Protestant' Ulster on the basis of a 3 to 1 nationalist majority in Ireland as a whole, while simultaneously insisting on a unionist right to rule Ulster with a 55 percent Protestant majority.[13]

This comment is also one-sided. The Nationalists were equally 'impervious to the logic of numbers', since the population distributions of Tyrone and Fermanagh, on which they were to base the case for exclusion, were almost as identically in their favour, as the distribution of population over the whole province favoured the Unionists.

So, according to the 'wishes of the people', as interpreted at this time, there was a case for excluding the whole province from the home rule scheme or, if county option were the basis, for excluding four; but there was no democratic justification for excluding six. As shown above, the Protestants had clear majorities in four counties, were a large minority in two, and a very small minority in three.

Carson was well aware of these demographic facts, but revealed his true intentions when he spoke at a gathering of southern Unionists on 10 October 1911, telling them that they need not fear desertion by their northern co-religionists; since 'if Ulster succeeded, home rule is dead'. It is hard to credit that so experienced a politician could really believe that a scheme for autonomy for the four provinces of Ireland could be frustrated by opposition from little more than half a province, especially since the majority within that province was largely concentrated in the counties east of the river Bann. Mansergh speculates that it was something Carson wished to believe and added that 'it proved a great miscalculation'.[14]

1912

The years 1912 to 1914 marked the most turbulent period in British constitutional history since the seventeenth century; a period when in order to oppose a modest measure of devolution in one part of the United Kingdom armed rebellion was threatened in another, together with a totally illegal alternative government, and these seditious schemes were strongly supported by the leader of the opposition in the House of Commons and his party. Historians tend to treat the sequence of events from the Lloyd George budget of 1909 to the passing of the

---

13 J.J. Lee, *Ireland, 1912–1985: Politics and Society* (Cambridge, 1989), p. 4. It must be remembered that the franchise was (until 1918) restricted to male householders and a few other limited male categories (lodgers and service voters).   **14** Mansergh, *The Unresolved Question*, p. 46.

Home Rule bill in June 1914 as a simple continuum, but there are important differences between the first two years and the last three. Opposition to the budget and the Parliament bill was strictly on constitutional lines, and resolutions on the Lords' veto and the passing into law of the Parliament bill took a mere fourteen months.

But the Home Rule bill was a different case. The monster meeting tactics continued to be used with ever greater efficiency. On Easter Tuesday 9 April 1912 a gathering of 100,000 at Balmoral on the outskirts of Belfast was addressed not only by Carson but also by the new leader of the Conservative Party, Bonar Law, who told the gathering that they held 'the pass for the Empire' and assured them that (they) would 'save the Empire by (their) example'.[15] (Redmond held a counter-rally in Dublin.)

While a study of these speeches and others during the succeeding years would lead invariably to the conclusion that the Liberal government were wholeheartedly committed to home rule for the whole of Ireland and that in consequence the Ulster Unionists would be backed by their colleagues in Great Britain in resisting even to the point of civil war, the private behaviour of some leading politicians belied their public rhetoric.

As has been said, Carson's first belligerent speech in Ulster was delivered on 23 September 1911 with its threat of a 'provisional government'. But, just one month previously, he had written to Curzon, showing his realization that an extremist policy might split the party, and urging that 'before much more time passes over we ought to know exactly what as a party we are prepared to do and the lengths to which we may consider ourselves entitled to go.'[16] On the other side, Churchill, who had jeered at Carson as the commander-in-chief of only half of Ulster and who made on 8 February 1912 a disastrous visit to Belfast (in which he was forced to take refuge in a Nationalist area), had just two days previously supported the first formal proposal in cabinet for separate treatment for Ulster.

The evidence of what transpired at the crucial cabinet meeting on 6 February 1912 is not entirely consistent. Patricia Jalland provides the most coherent account.[17]

It appears that Lloyd George proposed that the Protestant Ulster counties be allowed to 'contract out' of home rule. This was supported by Churchill, Haldane and Hobhouse, but vehemently opposed by the lord chancellor, Loreburn, and Morley – the position of other ministers is more difficult to establish – and voted down. Then a resolution, proposed by Crewe, was passed, rejecting exclusion, but reserving to ministers the right to alter (or drop) the bill, initially applicable to the whole of Ireland, if it should appear 'expedient' to provide special

15 *Belfast News Letter*, 10 April 1912.   16 Carson to Curzon, 21 August 1911, Curzon MSS EurF 112/18.   17 Patricia Jalland, *The Liberals and Ireland: The Ulster Question in British Politics to 1914* (Aldershot, 1993), p. 59, points out that Birrell himself in a letter to Churchill, as early as August 1911, suggested temporary Ulster exclusion by county option, but through fear of antagonising the Nationalists 'did not feel free to press it formally on the Cabinet or the public'.

treatment for the Ulster counties. It was also resolved that this decision be 'clearly' communicated to the Irish (that is, the Nationalist) leaders.

The attitude of Asquith is problematical. Hobhouse puts him among the supporters of exclusion, while his own letters indicate the opposite. Both accounts may be reconciled on the assumption that he first sided with Lloyd George and Churchill, then changed his mind as the discussion progressed. This would unhappily confirm the unflattering verdict of Charles Hobhouse on his leader: 'He is nearly always in favour of the last speaker, and I have never seen him put his back to the wall'.[28]

The question remains whether the cabinet decision was actually conveyed (as Mansergh assumed) to the leaders of the Irish Nationalists.[19] There is some evidence that Lloyd George met Dillon soon afterwards, but whatever was discussed does not merit a mention in the biographies of either Dillon or Redmond. Lloyd George's latest biographer suggests that it is likely that he did not propose to Dillon the outright exclusion of Ulster, as he had suggested in cabinet, but 'some special privileges for the North within the Dublin framework'.[20]

The history of the cabinet meeting of 6 February 1912 and the subsequent failure to follow it up underpins Patricia Jalland's thesis that Asquith missed 'one of the best opportunities history has ever offered for the peaceful solution of the Ulster problem'.[21] If only he had forced the Nationalists (in 1912) to accept what they actually conceded two years later, she claims, home rule with Ulster exclusion might have become a starting point for the peaceful evolution of an all-Irish state.[22]

This argument is unrealistic. There was no possibility that in the spring of 1912 Redmond and Dillon, buoyed up with the prospect of a parliament in Dublin, would agree to the exclusion of part of Ireland, and, even if they did, that they could carry their activists, particularly the Ulster Nationalists who were ridiculing the Ulster posturings.

Even if, uncharacteristically, Asquith were to compel Redmond – on whose support his parliamentary majority depended – to accept Ulster exclusion in 1912, this would not have destroyed the opposition to home rule. As Jenkins points out, half the British Conservatives, 'sick with office hunger', were simply playing the 'Orange card' in order to embarrass and hopefully overthrow the government. More important still was the opposition of the southern Unionists whose strength Jalland (whose researches end in 1914) greatly underestimates. Although after 1885 they never won as many as five out of the 103 Irish seats in the House of

18 Edward David (ed.), *Inside Asquith's Cabinet, from the diaries of Charles Hobhouse* (London, 1977), pp 111, 120.   **19** Mansergh, op. cit., p. 50.   **20** Bentley Brinkerhoff Gilbert, *David Lloyd George: A Political Life: The Organizer of Victory, 1912–16* (London, 1992), p. 94.   **21** Jalland, *The Liberals and Ireland*, p. 65.   **22** Ibid., pp 65–77. Incidentally the *Hobhouse Diaries* (p. 109) have two entries for December 1911 on the forthcoming Home Rule bill, but no reference to Ulster.

Commons, their influence was out of all proportion to their numbers. The great Irish landlords (Lansdowne, Midleton, Londonderry, Desart) sat in the Lords, and they could count on the support of the English landed families from the aristocratic Cecils to the squirearchy represented by Walter Long. The crucial question is that since, as we shall see, opposition from the southern Irish Unionists helped to wreck the settlement in 1916, which had secured the support of Carson and Redmond, in spite of patriotic appeals for national unity in the midst of war, is it at all likely that they would have been more accommodating in the peaceful conditions of 1912?

The third home rule bill – officially the Government of Ireland bill – was prepared by a cabinet committee which, after nearly four months' deliberation (mainly over the complex financial provisions) was introduced by the prime minister in the House of Commons on 11 April 1912. Its general character resembled that of the abortive Act of 1920. The first clause declared that the supremacy of the Westminster parliament remained 'unaffected and undiminished'. Certain services (for example, peace and war, coinage, the external postal services and the levying of new customs duties) were expressly reserved to Westminster. Control of the police was reserved for six years. The bicameral Irish parliament was debarred from either endowing religion or imposing religious disabilities. The taxing power was limited: until the existing deficit of £2 million of expenditure in Ireland over income was liquidated, the entire proceeds of all Irish taxes and customs and excise duties were to be paid into the British exchequer, from which in turn a block grant would be transferred annually to cover the transferred services. The imperial government was to control and pay for the reserved services.[23] The Irish parliament would be free to legislate on all other matters, subject, as had been said, to the overriding authority of the Westminster parliament. Ireland would continue to be represented at Westminster, but its numbers would be reduced from 103 to 42.

The parliamentary career of the home rule bill has been frequently summarized by many Irish historians and is exhaustively discussed by Patricia Jalland.[24] We shall concentrate on certain points that in our opinion have either been ignored or insufficiently emphasised by previous writers.

THE FIRST CIRCUIT (1912–13)

The first circuit began on 11 April 1912. The first reading lasted from 11 to 16 April. The second reading lasted intermittently from 30 April to 9 May. The committee stage began on 11 June but was still incomplete when parliament rose on

---

23 The complex financial provisions, devised by the young Postmaster General, Herbert Samuel, are discussed by Patricia Jalland in her PhD thesis. They took up much of the time of the Committee as Jalland states in *The Liberals and Ireland* '(the Samuel scheme) was so complicated that few people other than Samuel ever understood it' (p. 47). 24 Ibid., pp 80–120, 192–207, 248–20.

2 August. When parliament reassembled in October, a closure motion provided for 27 days for committee; seven for report and two for third reading. Accordingly, the committee stage ended on 12 December. The report stage was taken between 1 January and 13 January 1913; while the third reading occurred on 15 and 16 January.

Asquith's opening speech,[25] which a hostile observer said was 'by common consent altogether unequal to such an occasion',[26] apart from outlining the provisions of the bill, stressed the necessity for accommodating the deliberate constitutional demands expressed since 1885 by 80 per cent of the voters of the Irish nation, and at the same time of conciliating the minority by having a nominated senate. The supreme authority of the Westminster parliament remained unimpaired and the lord lieutenant would remain the head of the executive with a veto power over legislation. The only reference to the Ulster problem was an assertion that a small minority could not thwart the national will. He also hinted that Irish home rule would be 'the first step in a larger and more comprehensive policy'.[27]

Carson, who followed directly after Asquith, described the pledge of 'home rule all round' as 'utterly hypocritical'.[28] The only reason for the bill was the necessity to placate the Irish Nationalists, although in his opinion there were no outstanding Irish grievances. There was no suggestion of militant opposition from Ulster. Craig, following soon after, claimed that the demand for home rule had actually become less accentuated since 1893.[30] Redmond made an emollient speech, evoking (as had Asquith) the venerable memory of Gladstone introducing a similar bill twenty-six years previously. The remaining two days of the first reading debate were largely taken up with Conservative speeches. Balfour and Long derided the notion that the bill has a federal content and pointed out that the main Liberal federalist showpiece, South Africa, had evolved into a unitary state. Long was particularly bitter about the betrayal of the Irish Unionists generally who asked no more than to be allowed to continue, as they had ever been, 'loyal to the Crown and the flag'.[39] Bonar Law warned that 'you will not carry this bill without submitting it to the people of this country'.[31] Interestingly, none of the opposition leaders referred to the special problem of Ulster. Birrell, replying, accused Bonar Law of personal vituperation against Asquith. The first reading passed by 360 votes to 266.

The observation by Lloyd George in a letter to his mistress that the first reading debate showed 'no enthusiasm' on the Liberal side[32] (a judgment shared by Austen Chamberlain)[33] was confirmed when Churchill introduced the second reading in a speech, which may not have been his worst ever performance,[34] but

25 *Hans*, xxxii (11 April 1912), 1399–1426.   26 A. Chamberlain, *Politics from Inside*, p. 476.   27 Ibid., p. 1403.   28 'You have not the least intention of doing any such thing': ibid., p. 1433.   29 Ibid., p. 1474.   30 See J. Kendle, *Walter Long, Ireland and the Union, 1905–1920* (Glendale, CA, 1922), p. 71.   31 *Hans*, xxxvii, 16 April 1912, p. 301.   32 K.O. Morgan (ed.), *Lloyd George Family Letters* (London, 1973), p 161.   33 Chamberlain, *Politics from Inside*, p. 474.   34 Ibid., p. 480.

managed to avoid giving any exposition of the bill, or even an argument in support of it, apart from the merest generalities. Nevertheless, Churchill did admit that the opposition in Ulster might seriously frustrate government policy,[35] and the same argument was advanced by the foreign secretary, Sir Edward Grey, who pointedly replied to opposition jibes that the Liberals were dependent on the Irish Nationalists, by asserting that 'it is they who are dependent upon the Irish vote for any chance of turning us out'.[36] Grey stated that nothing but harm could come from denying that there were differences of national opinion within the United Kingdom. 'There is an Irish national feeling and there is national feeling in other parts of the United Kingdom. You cannot help it. The thing is there.' He was 'quite certain that the last thing an Irish Executive, or an Irish parliament, will do will be to provoke a strong minority in Ulster to resistance based upon moral wrong and unreasonable treatment'.[37] On the other side, Austen Chamberlain 'repudiated' the charge that 'Friends of mine on this bench' were 'inciting civil war in Ulster'.[38] The solitary Ulster Liberal, T.W. Russell, begged the house not to be 'deluded' by Unionist threats.[39]

Replying to the debate, Asquith again insisted that the constant demand by four-fifths of the Irish representatives could not be denied, brushed aside the opposition arguments, including the one that home rule was not an issue in the December 1910 election ('Not a single man on the opposition front bench … did not declare that home rule was an issue'),[40] and strongly challenged the opposition, 'Do they or do they not agree that if home rule is within the constitutional competence of this parliament with the approval of the electorate of the United Kingdom, Ulster is entitled to resist?'[41] There was no reply. The second reading was carried by 370 votes to 270.

However, a handful of Liberal MPs expressed doubts about the wisdom of trying to include Ulster in the bill, and one of them, Thomas Agar-Robartes, scion of an aristocratic Cornish family, known to hold strong Protestant opinions and a former Liberal-Unionist, tabled an amendment for the committee stage to exclude the four Protestant counties – Antrim, Down, Londonderry and Armagh – from the jurisdiction of the Irish parliament. This, the very first partitionist proposal in either house of parliament, took only three days of debate. Agar-Robartes claimed unconvincingly that 'everyone' would admit that Ireland consists of 'two nations, different in sentiment, character, history and religion'. By excluding 'the North-East Ulster nation' the chief danger of disturbance would be removed.[42]

35 *Hans*, xxxvii (30 April 1912), 1722.   36 *Hans*, xxxvii (2 May 1912), 2093.   37 *Hans*, xxxvii (2 May 1912), 2097.   38 *Hans*, xxxviii (7 May 1912), 265.   39 *Hans*, xxxviii (7 May 1912), 624. 40 *Hans*, xxxviii (9 May 1912), 695.   41 *Hans*, xxxviii (9 May 1912), 696.   42 *Hans*, xxxix (11 June 1912), 771–3. Isolated partitionist proposals had previously been made by Anthony Traille, the Antrim-born provost of Trinity College, Dublin, and by the maverick nationalist Arthur Clery: see A. Clery, *The Idea of a Nation*, ed. P. Maume (Dublin, 2002).

Birrell for the government argued (also unconvincingly) that far more evidence would be required to convince him that Ulster really wanted exclusion. The Ulster Unionists, who spoke tactically, favoured the amendment because they argued that without Ulster and its revenues home rule would be impracticable. Redmond, however, cited an article in the *Irish Times*, the leading organ of southern union-ism, which described the amendment as 'a trap designed to secure an admission that the Northern Unionists were willing to abandon the Unionists in the rest of Ireland to their fate'.[43] (Almost the same words were used by Walter Long, the former leader of the Irish Unionists, in a letter to Bonar Law, but he allowed him-self to be persuaded to support the amendment for tactical reasons.)[44] Redmond also rejected the two nations theory, claiming that the Irish nation included both Catholics and Protestants, and denying that even the four counties were homog-enous, with 315,000 Catholics to 725,000 Protestants.[45] An English-based south-ern Unionist, Walter Guinness, corrected him; the Protestant figure, according to the 1911 census, was 750,000.[46] Closing the debate, Lloyd George, in his only major contribution to the first circuit discussions, argued that the four counties were not a workable administrative unit and challenged the Unionists to prove their sincerity by putting forward their own proposals.[46a] This speech 'disarmed and confused the Government's critics, as was undoubtedly the intention',[47] but interestingly Lloyd George abstained from voting, as did 62 Liberals, while five voted with the opposition. The amendment was defeated by 320 to 251.

By the time the House adjourned for the summer recess virtually all the pos-sible arguments for and against the bill had been deployed. Broadly speaking, there were pragmatic arguments on both sides. Liberal spokesmen constantly adverted to the need to recognise the consistent demands of four-fifths of the Irish electorate and the subsidiary consideration that parliamentary efficiency would be enhanced if the Irish question were settled. Irish Nationalist speakers kept harping on the importance of acceding to the just demand of the Irish people, and frequently adverted to recent cases of autonomy being conferred on former colonies – Canada, Australia, as well as the special case of South Africa. Ulster Unionist speakers claimed that it would be an unprecedented injustice to cast a loyal population out of the United Kingdom and to subject them to a regime that would probably be hostile to their best interests. They also used the argument that the bill emerged not from any reason of principle but solely from a desire to placate the Irish Nationalists. The most complex set of arguments was advanced by British Conservatives. On the one hand they claimed that home rule would inevitably lead to separation and so destroy the integrity of the kingdom, on the

43 *Hans*, xxxix (13 June 1912), 1080.   **44** Kendle, *Walter Long, Ireland and the Union*, pp 72–3.
45 *Hans*, xxxix (13 June 1912), 1085–90, 1154–5.   **46** *Hans*, xxxix (18 June 1912), 1130.   **46a**
Hans, xxxix (18 June 1912), 15.   **47** Jalland, *The Liberals and Ireland*, p. 100. Jalland attributes
the failure to support the government variously to non-conformity, federalism and Scottish nation-alism: *The Liberals and Ireland*, p. 102.

other hand that Ireland's best interests, *pace* the nationalist pleadings, would be served by maintenance of the Union, and they pointed to the generous treatment by recent governments in the areas of land purchase, infrastructure, local government and education.

But the most intellectual Tories also used a theoretical argument to counter the Liberal claim that Ireland was a nation. L.S. Amery MP, Fellow of All Souls, in a widely read pamphlet, cleverly marshalled all the arguments, theoretical and practical, for his case.[48] The British Isles, he claimed, was racially a single and distinct area, with a single language, and a territory over which the Celtic and Teutonic elements were blended and intermingled in varying proportions. Historically the whole area had been under a single crown for 300 years and Ireland had been under the English crown for four centuries previously. To the argument that nationality is defined by a widespread demand for national separation, he replied that 'every argument that can justify one-fifteenth of the population of the United Kingdom in demanding separation from the United Kingdom is a stronger justification for one-quarter of Ireland in insisting that it shall not be governed from Dublin'. So the mere existence of Ulster destroyed the case for home rule and Amery was prepared to support Ulster resistance even to the point of civil war, although he did not believe that things would ever come to such a pass. Amery derided Asquith's claim that the 1912 bill marked a first instalment of federalism for the United Kingdom, firstly by claiming that the idea had not been carefully thought out, nor even seriously contemplated, and, more tellingly, he argued that the particular financial terms of the bill would not fit into a federalist model, because old age pensions and land purchase would be subsidized at the expense of the taxpayers of Great Britain. He also contended that the Irish parliament would regard a federal arrangement as the first constitutional step to independence, not the last. Another Tory intellectual, Balfour, later argued against the claim that Ireland was a nation on the ground that '(it) never had an organic political past as a single great community',[49] and that it possessed no surviving indigenous political institutions.

Reading these speeches and writings, one is impelled to the conclusion that these Tories imperfectly understood the concept of nationalism as it had evolved in nineteenth-century Europe. Amery's statement that every argument justifying the separation of Ireland from the United Kingdom could with equal force be used to justify the separation of Ulster from the rest of Ireland could be taken to imply what he sought to deny – that nationalist Ireland was indeed a separate nation (whether unionist Ulster was regarded as part of a wider British nationality or as a nation in its own right). The other geopolitical arguments could be countered by reciting the definition of a nation, as understood in both eastern and

48 L.S. Amery, *The Case against Home Rule* (London, 1912), p. 62.   **49** Shannon, *Balfour and Ireland*, p. 667.

western Europe at the beginning of the twentieth century as: 'the largest community which, when the chips are down, effectively commands men's loyalty, overriding the claims both of lesser communities within and those who cut across it, or potentially enfold it, within a still greater society'.[50] That nations emerging into statehood might have to make special arrangements for 'lesser communities within' is obvious and was recognised in some (though not all) new states created after the Treaty of Versailles.

In England at the turn of the century there was little interest in, or discussion of, the concept of nationalism. Not surprisingly, since apart from the miniscule Scottish and Welsh nationalist movements there was only the Irish nationalism to consider. However, in Europe nationalist movements had developed in the Austro-Hungarian, Russian and Ottoman empires and the writings of Mazzini, Kossuth and even lesser figures such as Pavlicek were popularized in Irish nationalist journals like *Sinn Féin*.[51] The theoretical arguments against home rule belong to the genre of anti-nationalism, or support for the *status quo*. There was in 1912 no claim for separate treatment for Ulster, but rather for keeping the province within the United Kingdom, and the economic argument was predominant.[52]

After the defeat of the Agar-Robartes amendment, there were two days in committee devoted to frivolous or drafting amendments proposed by Tories, including one deleting the provision for an Irish senate on the grounds that second chambers were unnecessary! The debate was adjourned on 1 July and did not resume until after the recess, on 10 October.

On Orangeman's day 1912 Bonar Law made a speech at Blenheim which denounced the Liberal government as a revolutionary committee that had subverted the Constitution, declaring that, 'there are things stronger than parliamentary majorities',[53] and pledging, 'I can imagine no length of resistance to which Ulster will go which I shall not be ready to support'.[54]

This inflammatory speech has passed into all the studies of the period. What has attracted little notice is the minor riot occurring in Belfast, sparked off by the Castledawson riot, at that very time. For about three weeks the shipyards of Harland and Wolff and Workman, Clark and Company, the flashpoints of earlier riots, were again convulsed, with the Protestant craft workers attacking the Catholic labourers.[55] Eighty people were injured, eight in danger of their lives.[56] Over the

50 R. Emerson, *From Empire to Nation* (Cambridge, Mass., 1960), pp 95–6. **51** C. O'Leary, *Celtic Nationalism* (Belfast, 1982). **52** The argument that Belfast and its environs had done particularly well out of the Union goes back at least to 1841. See I. Budge and Cornelius O'Leary, *Belfast: Approach to Crisis: A Study of Belfast Politics, 1613–1970* (London, 1973), pp 107–8; W. McComb, *The Repealer Repulsed*, ed. P. Maume (Dublin, 2003), pp xvi, 193–4. **53** *Hans*, xli (31 July 1912), 2128. **54** Stewart, *The Ulster Crisis*, p. 57. Robert Blake writes 'Such a tone had not been heard in England since the debates of the Long Parliament' (*The Unknown Prime Minister*, p. 130). **55** Budge and O'Leary, *Approach to Crisis*, pp 93–4. Catholics never accounted for more than 10% of the shipyard workers; see also, P. Bew, *Ideology and the Irish Question* (Oxford, 1994), pp 56–68; P. Maume, *The Long Gestation* (Dublin, 1999), pp 134–5. **56** The most savage incident reported

next two weeks, about 2,500 workers fled and eventually the management of Harland and Wolff issued a statement threatening to close down their shipyard.[57]

Devlin raised the matter in the House of Commons on 31 July 1912. In a forceful speech, relying only on reports in pro-Unionist, London newspapers, the 'pocket Demosthenes' denounced Law and Carson as the instigators of the riot. Referring to a speech delivered by Carson on 24 June in London in which he promised that, 'when he went to Ulster he intended to break every law that was possible', Devlin scornfully referred to Carson as, 'an academic anarchist' and ended, 'The only law that is broken is not the law broken by the ex-Solicitor-General for England but by his wretched dupes in the city of Belfast'.[58]

Ulster Unionist MPs (James and Charles Craig) expressed perfunctory regret for the disorders, but alleged that Devlin's charges were exaggerated, that he was provoking ill-feeling both in the House and in Belfast, and that anyway Protestant disorders were always in response to provocation from the other side. Bonar Law asserted that his Blenheim speech was carefully prepared, since he thought it 'quite possible' that 'many of my supporters in this House might think that I was going too far'.[59] But he claimed that the then situation was the most serious facing the country since 1642, and holding the views which he did hold he was bound to express them, and if any considerable number of MPs, or the party outside, disapproved, he would resign. Carson gave a short speech in a different tone. He ignored Devlin's taunts, claiming that he had never countenanced, and never would countenance, such actions as the 'lamentable' events in Belfast. He drew attention to a letter which he had sent to the Belfast papers appearing on 11 July, 'in the hope that any influence I possessed might be used towards preventing breaches of the peace upon the Twelfth'.[60] He ended, 'As far as I am concerned, I earnestly hope that peace may be restored in the shipyard and that there may be no further ground of complaint on the one side or the other.'[61] Birrell, speaking immediately afterwards, congratulated Carson on his speech.[62] By 2 August when parliament adjourned for the summer recess, the debate in committee had not got any further than the third clause.

Unquestionably the most spectacular events during the recess were the great demonstrations surrounding the signing of the 'Ulster Covenant' (19–28 September 1912) in which nearly half a million people throughout Ulster pledged themselves to use 'all necessary means' to defeat 'the present conspiracy to set up a home rule parliament in Ireland'.[63] Carson, who was feted like a triumphant

was that of a Catholic worker, stripped naked and held over a furnace, who was rescued by four men armed with sledge hammers threatening to smash the skulls of the perpertrators (*Daily Telegraph*, 5 July 1912), though the authenticity of this report was later questioned.   **57** Unlike his predecessor, the then controlling director of the yard, Lord Pirrie, had nationalist sympathies. The founder, Sir Edward Harland (who later served as a Unionist MP), had threatened to close the yard after the riots of 1864, but took no action after the more serious riots of 1886: Budge and O'Leary, *Approach to Crisis*, p. 94.   **58** *Hans*, xli (31 July 1912), 2096.   **59** Ibid., 2134.   **60** Ibid., 2108.   **61** Ibid., 2111.   **62** For the entire debate see *Hans.*, xli, 2090–134.   **63** A.T.Q. Stewart, *The Ulster Crisis*, pp 61–7.

general at a series of rallies working up to the mass signing on 'Ulster Day', made all his speeches on the same theme: there must be no home rule. Although visiting journalists were impressed by the demonstration, it did not impel them to the prediction that civil war was in the making, but rather that the crowds did not believe that they could possibly be beaten.

When parliament reassembled, the opposition were 'spoiling for a fight' and used the remaining days in committee obstructively until the government forced through a guillotine motion on 10 October.[64] This led to several wild scenes in the Commons, culminating in the celebrated assault on Churchill by an English-based Ulster Unionist MP, Ronald McNeill (13 November). The rest of the session was marked by protracted, but weary, resistance, and the committee stage was ended by 12 December.

The third reading debate, beginning on 1 January 1913, provided the last opportunity to amend the text of the bill. The Ulster Unionist Council, meeting in December, decided to table their own amendment, excluding the whole nine counties from the bill, and entrusted it to Carson. The cabinet meeting to discuss the Carson amendment (on 31 December) exhibited that curious indecisiveness that marked so much of its history. The prime minister insisted that the amendment was merely a wrecking device; the general view was that any concession at that stage would be fatal to the financial and other proposals of the bill and must be resisted. But Churchill and Lloyd George argued forcefully against 'banging the door against Ulster, [65] and 'ultimately it was agreed that the PM should speak rejecting the actual amendment but hinting that if settlement was in the air, our attitude would be one of meeting them half-way.[66] During the debate Carson alone argued for the exclusion of the whole of Ulster, giving interesting examples of regions with smaller area or population that had already achieved statehood.[67] The only ministerial speakers (Asquith and Churchill) argued on technical grounds; separation of the nine counties would make the rest of the bill unworkable. Apart from a few federally-minded Liberals, there were no speakers on the government benches and the amendment was easily defeated in a small House (294 votes to 197). The bill passed all stages in the Commons on 16 January 1913 by 367 votes to 257; two Liberals, including Agar-Robartes, voting with the opposition.

The bill then went to the Lords where it was given the same summary treatment as the bill of 1893 had received. The debate lasted a mere three days (27 to 30 January 1913), ending with a defeat on the second reading, by 326 votes to 69. No new point emerged during this brief debate. On the same day as the bill was rejected the Unionists lost a by-election in Londonderry city. This was

---

64 Stewart, op. cit., pp 64–5. Curiously the signing of the Covenant is omitted by Carson's latest biographer, Alvin Jackson.   **65** David (ed.), *Inside Asquith's Cabinet*, p. 127; Jalland, *The Liberals and Ireland*, p. 109.   **66** Jalland, op. cit., p. 109, omits this passage.   **67** 'Once again his purpose was chiefly, though perhaps not exclusively, tactical': Alvin Jackson, *Sir Edward Carson* (Dundalk, 1993), p. 31.

a gain by the Liberals (although virtually the whole Liberal vote came from Nationalists), who thereby doubled their representation in Ulster, and deprived the Unionists of their majority in the province. The new state of the representation was 16 Unionists, 15 Nationalists and two Liberals (the last being supporters of home rule). It is useful at this stage to examine the attitudes of the various parties at the end of 1912.

The Irish Nationalists, even in their private correspondence, regarded home rule as in the bag and ignored the dangerous potentialities of Ulster Unionism.

The Irish Unionists, both North and South, believed that their opposition, helped by their English allies, would be sufficient to kill home rule in any form.

The Liberal government supported Asquith's assumption that the whole bill could be pushed through in spite of Ulster opposition, although in cabinet three important members , Lloyd George, Churchill and Grey had serious misgivings.

As for the British Conservatives, their new leader, Bonar Law, had unequivocally supported Ulster resistance, believing that this could wreck the home rule scheme and force a dissolution, but his position was by no means secure. At this very time the party was racked by the long-standing internal struggle over 'tariff reform' – between free traders and supporters of taxes on food – which was coming to a climax. This conflict (which most Irish historians ignore but which encouraged nationalist belief that the Conservatives were too weak to resist home rule effectively) occupied all the attention of the party in the closing months of 1912.

Eventually after Law had threatened to resign, the party 'grass roots' (the National Union of Conservative and Unionist Associations) decided to abandon the policy of food taxes while remaining committed to tariffs on industrial goods – a decision which Law (a protectionist) reluctantly accepted (January 1913). He made the best of it by claiming that it was not a question of principle but of tactics! In Dutton's view, Law's handling of the whole crisis was 'strangely inept'.[68]

1913

The first half of 1913 was marked by three separate developments.

First came the second circuit of the home rule bill. The Parliament Act prescribed that a bill introduced under its terms in the second or third session must be presented in the identical form in which it had passed the Commons in the first session. Further Commons amendments were ruled out, but the Commons might 'suggest' further amendments to be considered by the Lords.[69] This in effect eliminated the committee and report stages during which Commons amendments were customarily moved and reduced the debates to two on the general principle of the bill – the second and third readings. The bill came back to the Commons for a considerably reduced circuit. The second reading took just two days and the third read-

68 Dutton, *'His Majesty's Loyal Opposition'*, p. 193.   **69** 1 and 2 George V (1911), 13.2.4.

ing one (7 July) in a much depleted House. The *new* feature was a Conservative demand for the bill to be submitted to the electorate: to this Asquith replied by inviting 'reasonable' suggestions from the opposition. The bill passed the final stage by 352 votes to 243 – due to pressure by the whips. After a two-day second reading, the Lords rejected the bill for the second time, by 302 votes to 64 (15 July).

The second development was the formation in January 1913 of the Ulster Volunteer Force which soon boasted the enrolment of 100,000 men – although the numbers have been disputed – under the command of a retired Indian Army officer, Lieutenant-General Sir George Richardson.[70] The standard view is that it was founded by Carson and Craig,[71] but although Carson attended numerous rallies there are important differences between the attitudes of the two men. In the previous month, Carson met thirty Ulster Unionist representatives in Belfast and apparently urged a compromise on home rule, but was voted down, while in May 1913, at a meeting of the Ulster Unionist Council, Carson 'gently' rejected a call for arms importation.[72] Throughout the summer and early autumn of 1913 when it looked as if the Home Rule bill would become law at the end of its third circuit in 1914, Unionist leaders took advantage of King George V's genuine concern that the Ulster resistance might get out of hand to suggest that the monarch refurbish rusty constitutional weapons to kill the bill, or at least force a dissolution on that issue. As early as March 1913 Bonar Law wrote, 'The difficulty is to find a method of securing it (an election) especially in view of the parliament Act. I hold that the position of the Sovereign is very important in this respect.'[73] What Law had in mind was indicated in an audience with the king, reported in a letter to Lansdowne (16 September 1913). He argued that 'undoubtedly while the king did not act, except on the advice of his ministers, his action was not purely automatic and if he has reason to believe that the advice was not in accordance with the wishes of "his" people on a matter of vital importance, he was entitled, under the Constitution, to dismiss the Cabinet and appoint other ministers who would advise a dissolution'.[74]

Law was recommending that the king dismiss his ministers and summon the leader of the opposition to form a government – a personal royal prerogative power last exercised by William IV in 1834.[75] Other Unionists even suggested that the king veto the bill when it would emerge from its third circuit – a power last exercised by Queen Anne in 1707. That proposal received little support, but a letter to *The Times* from George Cave, a Unionist lawyer MP (later lord chancellor) started a correspondence to which academic heavyweights as well as politicians contributed.[76] Cave argued that the monarch could dissolve parliament independently of ministerial advice. Sir William Anson, Warden of All Souls and a

---

70 See Jalland, *The Liberals and Ireland*, pp 132–6.   **71** Lee, *Ireland, 1912–1985*, p. 17.   **72** Jackson, *Carson*, p. 37.   **73** Law to A.V. Dicey (26 Mar. 1913), B.L.P., 33/5/20.   **74** Law to Lansdowne (16 September 1913), B.L.P., 33/5/56.   **75** Law had first made this suggestion to the king in September 1912 (see Jalland, *The Liberals and Ireland*, p. 130) and repeated it in a memorandum to the king on 31 July 1913.   **76** The correspondence in *The Times* (6–15 September 1913) is summa-

constitutional lawyer, supported Bonar Law's argument that the king could dismiss his ministers, if before taking action he could find an alternative cabinet to agree with him. Professor A.V. Dicey agreed that since 'the final decision of every grave question now belongs not to the House of Commons, but to the electors', the king would be justified in insisting on an appeal to the people. However, Professor J.H. Morgan argued that such a course would seriously compromise any future exercise of the normal power of dissolution.

Even before this correspondence, the king had demanded in a memorandum to the prime minister – in which he ruefully admitted that 'whatever I do I shall offend half the population' – that Asquith answer the Unionist arguments for the exercise of the personal prerogatives. Asquith who had a keen legal brain – as a young man he had drafted one of the great reformist measures of the nineteenth century, the Corrupt and Illegal Practices Act of 1883 – replied in a document 'of exceptional clarity and force' (September 1913).[77] He pointed out that the right of the crown to veto a bill, duly passed by both houses of parliament, had died in the reign of Queen Anne. Since then, although there had been monarchs of marked individuality and great authority none of them – not even George III, Queen Victoria or Edward VII, had ever dreamt of reviving this ancient prerogative. As to the power to dismiss ministers, Asquith conceded that it 'perhaps'[78] still existed, but recalled that when it was last exercised in 1834, when William IV 'one of the least wise of British monarchs'[79] dismissed the Melbourne cabinet, all that happened was that Peel took office with a minority government and asked for a dissolution. After the election the Whigs under Melbourne returned to power and the authority of the crown was damaged. Asquith concluded by pointing out that the Parliament Act, which dealt only with relations between the houses, had in no way changed the position, and urged that the crown avoid becoming 'the football of contending factions'.[80]

In another memorandum to the king Asquith conceded that if the home rule bill became law there was the prospect of bloodshed in Ulster, but to talk of civil war was 'a misuse of terms'. On the other hand, if it failed Ireland would become ungovernable. A general election before the bill became law would settle nothing. If the government won, the Ulster trouble would continue. If it lost, the problem of governing Ireland would become no easier.

King George replied, indicating that he was not convinced that the personal prerogatives in question were atrophied and strongly argued in favour of an appeal to the people before the bill became law.[81]

However, another suggestion by the king in early September of a conference between the party leaders to resolve the political deadlock did actually bear fruit

rized in W.I. Jennings, *Cabinet Government*, 3rd ed. (Cambridge, 1965), pp 539–45. **77** Roy Jenkins, *Asquith* (London, 1964), p. 283. For the full text see ibid., pp 543–5. **78** Jenkins, *Asquith*, p. 284. **79** Ibid. **80** Ibid. **81** Harold Nicolson, *King George V* (London, 1952), pp 222–4.

in an unexpected way. On 11 September 1913 an ex-lord chancellor, Lord Lore-burn, proposed in a letter to *The Times*, a conference or direct communication between the leaders 'in an attempt to reach agreement on the Irish question instead of fighting to the bitter end'.

The letter raised a storm of controversy. Loreburn, a life-long Liberal, had been in cabinet one of the strongest supporters of home rule, and Jenkins stern-ly rebukes him as one who 'with a typical elder statesman's show of non-parti-san wisdom, had embarrassed and irritated his former colleagues by writing to *The Times* to propound exactly the solution which he had so strongly opposed from inside'.[82] This is unfair, since Loreburn never mentioned Ulster exclusion, but the letter was widely misinterpreted as emanating from the cabinet instead of from an 'always disgruntled ex-colleague who in earlier days had said the Government should unhesitatingly root out disobedience',[83] and helped to create an atmosphere favourable to inter-party negotiations, as the king obviously wished.

Asquith did not welcome the Loreburn proposal. In a memorandum to the king, cited above, he wrote that the parties to this controversy, including Carson and Redmond, were not likely at that time to accept an invitation to discuss the government of Ireland, and that it was no good blinding one's eye to the obvi-ous fact that there was a deep and hitherto unbridgeable chasm of principle between the supporters and opponents of home rule.[84] However, the prime min-ister's hand was forced by the action of an impetuous colleague. Winston Churchill had not changed his opinion, first expressed in February 1912, that some form of Ulster exclusion might be necessary to save the bill. On meeting Bonar Law socially at Balmoral on 17 September he was delighted to find that Law was will-ing to consider Ulster exclusion as a basis for inter-party negotiations and urged secret talks between a few leaders as a preliminary to a wider conference. Churchill reported this conversation to Asquith, urging him to take the initiative suggest-ed by Bonar Law. (He did not mention that Law had added a condition that the exclusion of Ulster be not regarded as a betrayal by the southern Irish Union-ists.)[85]

Churchill followed this up with a speech in Dundee on 8 October 1913 in which he declared that the Unionists were then only claiming special treatment for Ulster instead of trying to block the entire home rule scheme, and that such a claim, if put forward sincerely, could not be ignored by the government. This was the first public admission by a cabinet minister that 'Ulster exclusion' might be an option.

Lloyd George, the other proponent of Ulster exclusion in 1912, was also active. On 29 September Lloyd George through a journalist intermediary passed to Bonar Law a suggestion that if the Unionist leaders in the Lords were to propose an

---

82 Jenkins, *Asquith*, p. 287, n.1.   **83** David (ed.), *Inside Asquith's Cabinet*, p. 147.   **84** See Jenkins, *Asquith*, Appendix B, p. 548.   **85** R. Blake, *The Unknown Prime Minister* (London, 1953), pp 150–9.

amendment involving the exclusion of Ulster, the government would then be able to enter a conference.[86]

On the same day as Churchill's speech Asquith yielded to pressure. In a rather stiff note to Bonar Law he acknowledged that Churchill had reported to him the substance of his conversation with Law, and that while a conference between the party leaders, as proposed by Loreburn, was out of the question, he understood that Law was proposing 'an informal conversation of a strictly confidential character' between the two leaders, and he (Asquith) would be happy to take part in such a conversation. [87] It was decided to hold the meeting at Cherkley Court, the estate of the young Unionist MP and newspaper proprietor, Sir Max Aitken, where security would be assured. Asquith rated security so highly that he did not inform any cabinet colleague except Crewe. (He also informed the king.)

The first of these encounters took place on 14 October.[88] Asquith found Law to be surprisingly frank. While acknowledging the prime minister's difficulty in securing the agreement of Redmond to any alternation in the home rule bill, he also stressed that he faced at least equally great difficulty if any inter-party arrangement would be regarded as a betrayal by the Southern Unionists. He admitted that he was personally in favour of home rule for the rest of Ireland with the exclusion of Ulster. Asquith was then emboldened to ask some direct questions. The first was, 'Are you prepared to throw the Unionist minority in the South and West to the wolves?' to which Law replied, 'Yes, unless we find that a majority, or at any rate a substantial minority, of the "sheep" protest.' Then Asquith asked what Law meant by 'Ulster'. Law evaded that question, which he said might be left over until later. Then, on the question whether the exclusion of Ulster would be permanent or temporary, Law replied that he was 'clearly in favour of an option of inclusion'. As to the form of government for the excluded areas, however defined, Law said he favoured a continuation of the *status quo*. As to whether Carson 'and his friends' would accept such a scheme, Law said that he thought that they would. For his part, Law recorded that to avoid misunderstanding he had asked Asquith what he understood the opposition's position to be. Asquith, after some discussion, replied, 'subject to the agreement of your colleagues whose concurrence is essential (the main party leaders), and if there is not a general outcry in the South and West of Ireland (i.e. among the Southern Unionists), if Ulster (which we can at present call X) were left out of the bill, then you would not feel bound to prevent the granting

---

86 See F. Harcourt Kitchin to Law (30 September, 1913), B.L.P., 30/2/35.   **87** The full text of the letter is given in Jenkins, *Asquith*, pp 287–8. Gilbert's claim that as early as August Asquith 'began to search through Lloyd George, for an avenue toward the settlement of the Ulster question' is at variance with Asquith's letters to the king and is not mentioned by Jenkins. See Gilbert, *Lloyd George*, p. 96.   **88** Of the two records of the meeting, a memorandum by Asquith, Asq. P. (xxxviii, ff 231–4) is more compressed than Bonar Law's letter to Lansdowne (15 Oct. 1913). B.L.P. 33/6/80.

of home rule to the rest of Ireland'. Law accepted that statement, and they agreed to meet again.

Between the first and second meetings of the two leaders, in spite of Asquith's desire for secrecy, reports circulated about the fact of the first meeting. This was due to Bonar Law, who felt it necessary to send a detailed report to Lansdowne, his chief lieutenant in the Lords, who had already expressed misgivings about the meeting and was willing only to consider complete exclusion of Ulster.[89] Law also corresponded with Balfour and they in turn communicated with other leaders, including Carson.[90]

So when the second meeting took place on 6 November,[91] both leaders were less optimistic than in October and they agreed at the outset that among the rank and file on both sides opinion was stiffening against compromise. On the question of having an election before the Home Rule bill could become law Asquith stated, and Law did not demur, that that would be the most dangerous expedient that could be risked, since if the Liberals won they would be compelled to go right on in spite of Ulster and that any settlement arrived at by the coercion of Ulster would be 'deplorable'.

Asquith then said that there were three possible ways of separate treatment for Ulster: (1) Grey's suggestion of 'home rule within home rule', (2) Ulster to be automatically included, and (3) exclusion until the people of Ulster wished to come into the Irish parliament.

Law indicated that only the third option was feasible. Then Asquith posed the question, what was meant by Ulster? Law asserted that Carson always spoke of the whole province, but Asquith said that it would be impossible to include the overwhelmingly Catholic counties of Donegal, Monaghan and Cavan. Law replied that he thought the minimum that Carson would accept would be the six counties.

When Law claimed that such a settlement would be impossible if the Nationalists were opposed, Asquith replied that 'Redmond and Co.' must choose between such a settlement and nothing.

When Law raised the question of administrative changes following on home rule, Asquith brushed it aside as of secondary importance. Law clearly saw in the course of the discussion that 'two things would be essential for him and must be granted by us if a settlement were to result'. The first was that Lansdowne would give an undertaking to do his best to get the Lords to allow the bill through and secondly that after the bill became law it should get a fair trial and would not be altered, even if the Conservatives were to win the following election.

89 Jalland, *The Liberals and Ireland*, p. 153.   **90** Balfour informed Law that if Asquith offered the exclusion of the four predominantly Protestant counties, the Unionists would be in some difficultly if they refused. Balfour to Law (17 Oct. 1913), cited in Shannon, *Balfour and Ireland*, p. 185, n. 58. **91** Again, the primary sources are pencilled notes by Asquith, recorded in extenso in Jenkins, *Asquith*, pp 290–2, and a letter from Bonar Law to Balfour (7 Nov. 1913), B.L.P. 33/6/93. The two accounts

Finally Law asked what was the next step. Asquith then promised definitely to bring his proposal before the cabinet at its next meeting on the following Tuesday.

In his covering letter to Balfour[92] and in another letter written on the same day to Walter Long,[93] Law admitted that he had feared that Asquith would demand a conditional acceptance from the Conservatives; but from the fact that he had not done so Law concluded that he was really serious about a settlement as the only alternative to an election. If he could not bring the Nationalists in, 'he will be in a bad way'. In a letter to Long, Bonar Law stated that Asquith had told him definitely that he would propose to the cabinet the exclusion of either the four, or the six, counties, 'probably the six'.[94] In reply, Long warned Bonar Law that while he would support any decision aimed at between Law and Lansdowne, he believed that any arrangement with the government would run the risk of splitting, and even of smashing, the Conservative party.[95]

There were two cabinet meetings on 12 and 13 November, after the second of which Asquith reported to the king. The prime minister reported, as he had promised, on his second conversation with Bonar Law, but 'thought better of putting this plan to his colleagues' and asked Lloyd George to put it forward as his own proposal.[96] The chancellor proposed, as a basis for possible compromise the exclusion of 'the Protestant counties' for a definite term of five or six years, with automatic inclusion at the end. This course, he claimed, would have two advantages: no one could support or sympathize with the violent resistance of Ulster to a change which would not affect them for years to come, and before the automatic inclusion took place, there would be two general elections, which would give the British electorate – with experience of the actual working of home rule in the rest of Ireland – the opportunity of continuing the exclusion of Ulster.

Asquith reported to the king that his suggestion, 'met with a good deal of support', but mentioned Birrell as strongly opposed to a 'bad and unworkable expedient'. (He also stated that the Liberal rank and file was increasingly opposed to compromise, because they 'wholly disbelieve in the reality of the Ulster threats'.)[97]

That the cabinet was divided on the Ulster issue is beyond question. Four separate groups may tentatively be identified. First the pro-Ulster faction comprising three powerful ministers, Lloyd George, Churchill and Grey. Then the 'ministerial die-hards', who were consistently against Ulster exclusion, were led by Reginald McKenna (who may have been influenced by his profound antipathy to Lloyd George and Churchill) and comprised the radical group minus Lloyd George; Walter Runciman, Lewis Harcourt, Herbert Samuel and John Burns.

---

are here collated.   **92** Law to Balfour (7 Nov. 1913), B.L.P., 33/6/93.   **93** Law to Long (7 Nov. 1913), B.L.P., 33/6/94.   **94** Ibid., B.L.P., 33/6/94.   **95** Long to Law (9 Nov. 1913), B.L.P. 30/4/11.   **96** Gilbert, *Lloyd George*, p. 97.   **97** Asquith to the king (14 Nov. 1913), Asquith Papers, vii, ff 71–2.

Thirdly, the 'waverers', Asquith, Crewe, Haldane and Morley, who had softened his earlier attitude through fear of bloodshed. Lastly, the remaining eight members of the cabinet, who appear to have had no strong opinions one way or the other but followed the majority.[98]

The cabinet meeting on 13 November resolved that Asquith would discuss its 'suggestion' with John Redmond on the following Monday. Like Dillon in Ballaghaderrin, Redmond had seen no reason to leave his Wicklow mountains residence during the summer of 1913, despite warnings from T.P. O'Connor in London. So he was alarmed by Asquith's suggestion. Redmond told Asquith that he could conceive of no other proposal which would create such a compact and hostile body of Irish opinion, both Unionist and Nationalist. Only if proposed by Bonar Law at the last moment as the price of an agreed settlement, would he consider it. But he would be prepared to concede to Ulster 'home rule within home rule'.

On 24 November Redmond sent Asquith a long letter in which he skilfully marshalled every possible argument against the government proposal.[99] Ulster exclusion would be bitterly opposed by Irish Unionists as well as Nationalists – he cited the *Irish Times* denunciation of it as the worst form of separatism. Even the English Unionists were not universally for it – he mentioned Lord Lansdowne – and it would split both the Liberals and Nationalists. Any such proposal by the government would be interpreted as a victory for the 'Orange threats'. Even at this point Redmond argued that the danger from Ulster was 'considerably exaggerated', and urged the government not to make any proposals but to wait and let them come from Bonar Law when the passage of the bill would be clearly certain.

The cabinet discussed Redmond's letter at a long meeting on 25 November. In a guarded reply Asquith assured Redmond that while there was no question of an offer to Bonar Law at that time, the cabinet must be free, 'when the critical stage of the bill is ultimately reached' to do what it thought best.[1] However, at the same meeting Grey proposed that in any subsequent conversation Bonar Law should be told that the Liberal Party could not agree to the permanent exclusion of Ulster, but would be prepared to discuss separate administrative arrangements or temporary exclusion.

Meanwhile, Lloyd George and Churchill again got into the act. Lloyd George met Redmond on 25 November and tried to get his agreement to the Lloyd George proposal by threatening that if no offer were made to the Tories, Grey, Haldane, Churchill and possibly himself might resign. However, when Redmond met Bir-

98 Jalland, *The Liberals and Ireland*, pp 58, 168, includes Birrell among the waverers, but, whatever his private misgivings might have been, Birrell had consistently supported the entire home rule scheme in cabinet discussions up to this time. Indeed, he felt so strongly about the Cabinet decision of 13 November that on the same day he sent a letter of resignation, which Asquith rejected. **99** Redmond to Asquith (24 Nov. 1913), Asq. P., xxxix, ff 29–35. **1** Asquith to Redmond (26 Nov. 1913), Asq. P., xxxix, ff. 36–7.

rell two days later, he was informed that Lloyd George's assertions were ridiculous since the cabinet had never even considered the proposals – which Asquith had informed the king had met with a good deal of support![2] Not surprisingly, these confusing signals left Redmond both angry and distrustful towards the cabinet and especially the prime minister.

Similar results followed from a conversation between Churchill and Austen Chamberlain on 26 November. Churchill told Chamberlain that the cabinet had never excluded the possibility of separate treatment for Ulster. Chamberlain's reaction was that while to him the bill without Ulster was only one degree less bad than the bill with Ulster, the only basis for a compromise was the course urged by Bonar Law in the second of his conversations with Asquith – optional inclusion by plebiscite after a fixed period. Churchill then pointed out that Asquith could not make any advances at that time since public opinion was not ready, and that Chamberlain should remember that 'we think that time is on our side'.[3] Chamberlain was left with the impression that while Churchill, Lloyd George, Grey and Asquith genuinely desired a settlement, they had no clear ideas as to the means and he believed that Asquith would 'wait and see' until the last moment.

On 10 December, following pressure from the king, the prime minister held the last of the three 'conversations' with Bonar Law. The meeting was so fruitless that Law wondered why Asquith had taken the trouble of seeing him at all.[4] Asquith stated that the proposal favoured by the cabinet was Ulster exclusion for a fixed period, followed by automatic inclusion. Law instantly dismissed it as totally unacceptable to the Ulster Unionists. When Asquith proposed the Grey solution, Law reminded him that the essence of Ulster's grievance was that they wanted to be treated like other citizens of the United Kingdom. Finally Asquith promised to carefully consider Law's remarks, but Law believed that Asquith's intention was simply to let things drift.

Before the end of the year Asquith made one more effort by inviting Carson to a secret meeting. Carson, who in private was much more conciliatory than his public *persona* would suggest, had already, in a much quoted letter, stated that if home rule was inevitable and the six Plantation counties were to be excluded it would be the 'best settlement', even if it meant the abandonment of the Unionists of the South and West. ('I have such a horror of what may happen if the bill is passed as it stands … that I am fully conscious of the duty there is to try and come to some terms.')[5] Having consulted Bonar Law, Carson agreed to meet Asquith. In fact two meetings were held. Asquith unexpectedly made fresh proposals of his own, which he termed 'veiled exclusion',[6] that an undefined area, termed 'statutory Ulster', should be given special powers of veto within the Irish

2 Asq. P., vii, ff. 77–8.   3 Memo of conversation with Winston Churchill (27 Nov. 1913), A.C.P., 11/1/21.   4 Law to Lansdowne (10 Dec. 1913), B.L.P. 33/6/111.   5 Carson to Law (20 Sept. 1913), B.L.P. 30/2/15.   6 Memo by Asquith (2 Jan. 1914), Asq. P., xxxix, ff. 72–4.

parliament through a majority of its members and should have limited administrative powers within its jurisdiction. Carson dismissed the suggestion out of hand and refrained from producing any alternative scheme. Shortly afterwards Bonar Law announced on 15 January in a speech at Bristol that the 'secret' negotiations were at an end.

By the end of 1913 the leaders of three of the five main groups had moved beyond their position of a year previously.

The Liberal leader was now prepared to consider some form of separate treatment for Ulster, although he was not clear as to what was feasible; the Conservative leader was also prepared for Ulster exclusion, even if it meant deserting the Southern Unionists; the same sentiments were shared by the Ulster Unionist leader. The Irish Nationalist leader had at last woken up to the danger from Ulster, but was confused and annoyed by the tactics of the British government. However, the southern Irish Unionists remained convinced not only of the righteousness of their objections to home rule, but of the ultimate success of their cause.

1914

The year 1914 began with an ominous escalation of the resistance in Ulster. Elements in the Ulster Volunteer Force were growing tired of practising with dummy rifles and pressed the leadership to allow the importation of real arms. At a meeting in Belfast on 20/21 January Carson reluctantly agreed to a proposal that guns and ammunition be imported from Germany. Meanwhile the government was warned that 95, 000 rounds of ammunition had been discovered in Belfast.

Carson's 'momentous'[7] decision to countenance the illegal importation of arms meant that he was then moving perilously close to treason. For the next two months there was a real possibility that Ulster Unionist leaders might be arrested and indicted.

At the same time, under the influence of Lord Milner (who at the end of 1913 had suddenly expressed a great interest in the Ulster cause) and L.S. Amery, a British Covenant on the same lines as the Ulster Covenant of 1912 was launched, initially signed by several peers of the realm and various academics, and ultimately by two million people.[8] The dissident Nationalist MP, T.M. Healy, was informed by the nephew of an English peer that 'there would be civil war in England as well as in Ireland, and that [Lord] Willoughby de Broke and his men would ride up to London and attack Asquith, and that the soldiers would not resist.'![9] Finally to compound the anxieties of the government (and Redmond) a body called the Irish Volunteers, obviously modelled on the Ulster Volunteers, had come into existence in November 1913.

---

7 Jackson, *Carson*, p. 38.   8 Stewart, *The Ulster Crisis*, pp 130–5.   9 Ibid., p. 133.

Faced with all this, Asquith once more exhibited his characteristic dilatoriness. At a cabinet meeting to discuss the forthcoming parliamentary session on 22 January he proposed (and they accepted) 'home rule within home rule' – an Ulster veto on any act of the Dublin parliament affecting the Protestant counties – although he knew Carson had already rejected it. As Jalland surmises, 'Some ministers were probably too preoccupied with other matters, or too ignorant of the detailed development of the Irish negotiations, to question the value of publicising a scheme which had been totally rejected by the Opposition.'[10] Others (especially Lloyd George and Churchill) probably acquiesced because they regarded the offer, when publicized (as the cabinet had agreed) as likely to demonstrate the willingness of the government to behave reasonably towards Ulster.

Before that could happen, however, the prime minister received another peremptory letter from Windsor Castle. Following the public admission that the inter-party talks had broken down, Bonar Law had urged the king to send a formal letter to Asquith, demanding a dissolution before the home rule bill became law. George V did not do this, but advised his prime minister that, irrespective of the guarantees offered, Ulster would never agree to come under the jurisdiction of a Dublin parliament, and warned that he regarded it as his duty to prevent civil war in Ulster.[11] Consequently, when Asquith met Redmond on 2 February to communicate the cabinet decision on home rule within home rule, he also mentioned the royal anxieties and the distinct possibility that the king might dismiss his cabinet and by appointing Bonar Law as prime minister precipitate an election. He also told Redmond that the Conservative majority in the Lords might reject the annual Army bill which was due for renewal in February or March.

In a long and dignified letter on 4 February, [12] Redmond insisted that he and his colleagues regarded the risk of civil war as 'greatly exaggerated' and urged that if the government were to announce any concessions to Ulster at the outset of the third circuit debate, this would be regarded as a victory by the 'Orange Party' and a betrayal of the Nationalist cause, and moreover – as should have been obvious – they would not satisfy the Unionists.

Finally, he urged the prime minister to limit his statement in the Commons to 'a reiteration of your frequent declarations that you were ready to consider favourably all proposals consistent with an Irish parliament, an Irish Executive, and the integrity of Ireland'.[13] Redmond's reply so impressed the cabinet that they decided that Asquith should make no offer of concessions at the opening of the parliamentary session – to the surprise of the opposition. But soon Lloyd George resurrected his scheme of county option by plebiscite, and home rule within home rule was quietly forgotten. Lloyd George argued that his scheme would have two essential, tactical advantages: it would be an offer the rejection of which would put the

10 Jalland, *The Liberals and Ireland*, p. 190.   **11** King to Asquith (26 Jan. 1914), Asq. P., xxxix, ff. 97–8.   **12** Asq. P., xxxix, ff. 81–4.   **13** Asq. P., xxxix, ff. 115–16.

Unionists entirely in the wrong as far as the British public was concerned, and it would not involve any alteration in the scheme of the bill.[14]

After an exchange of memoranda two meetings (27 February and 2 March) were held between Liberal ministers (Asquith, Lloyd George and Birrell) and Nationalist leaders (Redmond, Dillon, Devlin and O'Connor) at which the Irish very reluctantly accepted the proposal that individual Ulster counties might by plebiscite opt out of home rule for up to three years, followed by automatic inclusion without further legislation.[15] In the meantime the excluded area would continue to be governed by Westminster.

In a letter to Asquith on the same day summarising their decision, Redmond claimed that although his party by accepting even temporary partition would run 'enormous risks' of repudiation by their electorate, nevertheless he would accept the proposal of optional exclusion as the 'price of peace' but would insist as a condition of their assent that this scheme would be the last word of the government, that it be put to the opposition as one which the government could not under any circumstances enlarge and that if the opposition rejected it the government would pass the bill as it stood. Redmond also insisted that the Irish party not be required to vote for it.[16]

The cabinet considered Redmond's reply on 4 March and were prepared to accept his conditions. It was agreed that Asquith would unveil the county option scheme, when the debate on the third circuit of the Home Rule bill began on 9 March. Unfortunately the cabinet decision was leaked to the lobby correspondent of the *Daily News,* who published it on 5 March. This caused an immediate outcry among the Unionists, who insisted that a three-year exclusion period would be far too short. The same view was expressed by the king, who wrote that it could not be accepted by Carson.[17] So Asquith went back on his 'last word' and sent Birrell to present Redmond with an extension of the exclusion period to five years. On the following day Asquith demanded a further extension to six years to allow for *two* Westminster elections before the exclusion period would have expired. While expressing his 'deepest disappointment',[18] Redmond agreed to the latest proposal.

Redmond's position was obviously weakened by these enforced concessions, but, somewhat to his surprise, Devlin and another Ulster MP secured 'eminently satisfactory' results, when they consulted Nationalist activists in Ulster, largely through the influence of Cardinal Logue and Bishop O'Donnell of Raphoe. Their case was helped by a verbal promise from Asquith that in addition to the counties the boroughs of Derry and Newry with their large Catholic majorities should enjoy the benefits of local option.[19] Also the West Ulster Nationalists were

14 Lloyd George to Asquith (14 Feb. 1914), Asq. P., xxxix, f. 120.   **15** Lyons, *John Dillon,* p. 347.
16 Asq. P., xxxix, ff. 135–41.   **17** King to Asquith (4 Mar. 1914), Asq. P., xxxix, ff. 143–4.   **18** Redmond to Asquith (6 Mar. 1914), Asq. P., xxxix, ff. 145–6.   **19** Eamon Phoenix, *Northern Nation-*

convinced that county option would result in their coming under the Dublin parliament.

The final second reading of the Home Rule bill was scheduled for 9 March. The cabinet had to decide between the only two methods for passing the county option scheme simultaneously with the original bill. It could be presented as a 'suggestion', but this would incur the risk of being mauled by the opposition in the Commons, or by the Lords. The alternative was to incorporate the scheme in a second, amending bill to be simultaneously introduced. The cabinet believed that the result would be that the four Protestant counties, together with the county boroughs of Belfast and Derry would opt out of the scheme.[20]

Asquith presented his proposals when opening the second reading debate on 9 March. In a low-key speech he only gave the barest outline of the scheme, but acknowledged that the reason for its introduction was to avoid 'the dangers which lie ahead'. 'On the one hand, if home rule as embodied in this bill is carried now, there is ... in Ulster the prospect of acute dissension and even of civil strife. On the other hand, if at this stage, home rule were to be shipwrecked, or permanently mutilated, or indefinitely postponed, there is in Ireland, as a whole, at least an equally formidable outlook'.[21]

However, the Opposition proved intractable. Carson in a memorable phrase stated that Ulster did not want a sentence of death with a stay of execution for six years, while Bonar Law offered a counter proposal, a referendum over the whole United Kingdom, which he must have known that Asquith would never accept, since it was tantamount to the old demand for a general election on the home rule issue.

In the succeeding week Asquith refused all opposition demands to specify his intentions and on 16 March they tabled a no-confidence motion, which was debated on 19 March. This was the occasion when Carson stalked out of the House saying that he was returning to his 'own people'.[22] It was widely expected that when in Belfast he would declare the formation of a provisional government, but he did not do so. He was back in the Commons within ten days.

Meanwhile the so-called 'Curragh Mutiny' took place between 20 and 23 March. This story of monumental official bungling and incompetence has often been told and may be summarised here. A special cabinet committee was established on 11 March to deal with the dangers posed by the UVF – then over 100,000 in strength and allegedly equipped with 80,000 rifles. The committee decided as a precautionary measure to order the commander-in-chief in Ireland, General Sir Arthur Paget, to send troops to protect various arms depots in Ulster. Unfortunately, while Paget received his (oral) instructions from J.E.B. Seely (secretary for war), they were combined with an instruction that officers domiciled in Ulster might be excluded from the orders to move.

Thoroughly confused, Paget presented the instructions to his commanders in such a way as to give them the option of doing their duty, or facing dismissal. Of the two most senior officers, Major-General Sir Charles Fergusson, persuaded his officers to obey orders, but Brigadier-General Hubert Gough (an Ulsterman), commander of a cavalry brigade, and some 60 of his officers resigned.

In the farcical aftermath of the 'mutiny' Gough was summoned to the War Office, and on being invited to return to duty managed to extract from Seely and General French (the CIGS) a written pledge that the army would not be required to enforce the home rule Act on Ulster. This pledge was taken to the cabinet room and endorsed by the erratic John Morley. Too late, Asquith realized what had happened and on 25 March publicly repudiated the pledge. Seely, French and another general resigned, and Asquith took over the War Office.

Jalland asserts that 'The Liberal Government's home rule policy was undermined far more effectively by the Curragh crisis than by the intervention of the First World War'.[23] This is an exaggeration. In the first place no officer disobeyed an order. The subsequent history of the suppression of the Easter Week rising and the 'Troubles' of 1919–21 do not reveal a single officer of Irish origin refusing to obey an order to suppress the rebellions. It is hard to believe that if the UVF had attacked the army in 1914, Ulster-born officers would have behaved any differently. Summing up the 'Curragh incident', the son of Major-General Fergusson, the most level-headed of the senior Irish commanders, wrote: 'In fact orders given were, as Seely told the House of Commons on the Monday, 'punctually and implicitly obeyed', and all survivors of the Cavalry Brigade agree that had they received similar orders, straightforward and without hypothetical questions, they would have carried them out likewise.'[24] The opposition did not make much capital out of the 'mutiny'. In the succeeding weeks they tried to prove that there had been a 'plot' by certain ministers, especially Churchill, to provoke the Ulster loyalists into violence. Churchill easily refuted such charges by arguing that the government had the right to suppress armed rebellion by force and the Curragh incident actually strengthened support for home rule among grassroots Liberals who disliked being threatened. The Conservatives were also embarrassed by a fierce attack from the Labour benches for fomenting discontent in the army; trade unionists asked whether similar leniency would be shown to soldiers refusing to repress strikers.

The prime minister's decision to take over the War Office in succession to Seely had a good effect on the army, since, according to the leading historian of

---

23 Jalland, *The Liberals and Ireland*, p. 247.   **24** Sir James Fergusson, *The Curragh Incident* (London, 1964), p. 200.

the 'mutiny', 'the soldiers trusted Asquith'.[25] But under the archaic electoral law, which prevailed until 1918, the acceptance of any cabinet office necessitated a by-election in the minister's constituency. So Asquith had to travel to East Fife. His return was unopposed, but he was obliged to be absent from Westminster from 30 March until 14 April.[25a]

During Asquith's absence the Commons experienced a remarkable change of mood from the fierce dissension of the previous weeks. The adjourned debate on the second reading of the Home Rule bill was resumed on 31 March. Many back-benchers contributed as well as party leaders to a debate, which was marked by moderation on both sides.[26] Commenting on the debate on the following day, the usually abrasive, John Dillon stated that although he had been present at all debates on home rule over thirty years, 'yesterday for the first time I heard this question debated in a spirit of reasonableness and conciliation and with an evident desire on both sides to reach a settlement'.[27] On his return Asquith was greeted with another royal letter urging him to persuade Redmond 'for the sake of peace' to agree to permit the six counties to contract out without any plebiscite, and to remain out until such time as parliament decided otherwise.[28] However, Asquith rightly believed that Redmond had already been pushed to the limit of his endurance. The king's letter reminded Asquith that he had promised to renew his 'conversations' with Bonar Law and Carson, but before any further meeting could be arranged, there occurred the audacious gun-running at Larne,[29] involving the unloading of 35,000 rifles under the noses of the police (24–25 April).

The reaction of the cabinet to this blatant illegality was, to use a modern metaphor, of the 'headless chicken' variety. On Monday 27 April they ordered the prosecution of Carson and his leading supporters, but decided to begin the prosecutions by a method which did not involve arresting him – 'exhibiting an information' in Dublin; on Wednesday they sent a gunboat to sail up and down the Ulster coast in case there might be another gun-running; on Friday, having thought things over, they decided not to prosecute at all.[30] In fairness, a factor in that decision was advice from Birrell and Redmond that prosecution would convert the Ulster Unionist leaders from rebels into martyrs.[31]

Jalland writes that Carson was far more intransigent after the Curragh incident than in the previous year.[32] This is a superficial view. Stewart records that although pressed by his Ulster colleagues to set up the provisional government, throughout May and June, Carson 'had kept his cards very close to his chest'.[33]

25 A.P. Ryan, *Mutiny at the Curragh* (London, 1956), p. 160.   **25a** One cabinet minister wrote of 'the consummate genial manner of our great Prime Minister who never lost his temper and kept his head, and his judgement cool and collected throughout an exciting period', Pease diaries (27–31 March 1914), f. 88.   **26** *Hans.* lix (31 Mar. 1914), 1038–114. Carson was present, but did not make a speech.   **27** *Hans.* lix (1 Apr. 1914), 1204.   **28** King to Asquith (7 Apr. 1914), Asq. P., xxxix, ff. 157–8.   **29** Jenkins, *Asquith*, pp 314–16; Lyons, John Dillon, p. 349.   **30** Pease Diaries, ii, fo. 89.   **31** Lyons, *John Dillon*, pp 349–50.   **32** Jalland, *The Liberals and Ireland*, p. 249.   **33** Stewart, *The*

In fact Carson was deeply worried. He could see that his great objective of using the Ulster cause to defeat home rule would not be attained; he was afraid that he could no longer control his rebellious followers and that the civil war, which in his heart he dreaded, would be unavoidable.

So in the month of May 1914, 'King Carson', the leader of the Ulster rebels, actually seriously considered a federal all-Ireland solution in private discussions with the former Irish Unionist MP, Sir Horace Plunkett. Interviewed by Plunkett in January, 1914, Carson 'confessed his incapacity to control his own forces. He does not at all agree with their view'.[34] In April Plunkett wrote that 'He wants a settlement. He recognises the awful responsibilities attaching to his own part in leading the resistance of Ulster'.[35] In May, Carson tentatively proposed the idea of a federal solution for Ireland, but secured no support from his party;[36] but as late as June 1914 he was, 'looking for a solution along these lines – separate administration for Ulster in a way which could be continued in a united Ireland'. Plunkett added: 'This is a hopeful sign'.[37]

Although nothing came of the federal scheme, these exchanges demonstrate that Carson was genuinely anxious for a solution in spite of all the difficulties.

The difficulties were compounded by the actions of the government. On 5 May, prompted by the king, Asquith re-opened negotiations with Bonar Law and Carson. At this meeting Carson is on record as saying: 'Only a fool would fight if there is a hope of accommodation, and what a great thing it would be if this long-standing controversy could be settled once and for all.'[38] Some progress was made: it was agreed that a committee stage for the Home Rule bill could be dispensed with and that a separate amending bill should receive the royal assent on the same day as the original bill. But they could not agree on the contents of the amending bill.

On 12 May the prime minister announced to the Commons that he hoped to secure the third reading of the Home Rule bill before the Whitsun recess and promised that the amending bill would be introduced in the Lords.[39]

During the third reading debate the Conservatives, exasperated by their failure to extract from Asquith any details of the amending bill, created such an uproar that the Speaker had to adjourn the House.[40] When the debate was resumed, Asquith in a conciliatory speech revealed that the amending bill would embody the same proposals as his speech of 8 March, that is, county option for six years. On the same day the bill passed its final stage by 351 votes to 274.[41]

The amending bill was scheduled for the House of Lords on 23 June. In the meantime the government was faced with another worrying development in Ire-

Ulster Crisis, p. 221.   **34** Plunkett Diary, 17 Jan. 1914.   **35** Plunkett Diary, 1 Apr. 1914.   **36** *Belfast News Letter*, 1 May and 7 May 1914.   **37** Plunkett Diary, 19 June 1914.   **38** Stewart, *The Ulster Crisis*, p. 215.   **39** *Hans.* xlii (12 May 1914), 955.   **40** *Hans.* xlii (21 May 1914), 2181–214.   **41** *Hans.* xliii (25 May 1914), 80–94.

land.[42] Up to the end of 1913 the Irish Volunteers had fewer than 2,000 members, but by the summer of 1914 their estimated numbers had shot up to 160,000 – one third in Ulster. The Curragh mutiny had encouraged the view that nationalists needed a military force of their own. So there were now two paramilitary forces in Ireland, and the potential for armed conflict had obviously increased. Seeing this, the Nationalist leaders determined to bring the movement under the control of the Parliamentary Party, and after some desultory negotiations with Eoin MacNeill (the volunteers' chief of staff), whom he found 'vague and exasperating' – Redmond took the drastic step of issuing a public statement explaining that he must control the volunteers, or take steps which might split the movement. (9 June). The provisional committee of the volunteers yielded, with a bad grace. Lyons asserts that Redmond's position was 'very strong'[43] at this time, but his peremptory action caused great resentment among the volunteer committee and paved the way for the split (which occurred later in the year) and eventually for the Easter rising.

On 23 June the marquis of Crewe (secretary of state for India and leader of the House of Lords) introduced the amending bill, allowing each Ulster county to opt out of home rule by plebiscite for a period of six years.[44] If, he said, their lordships were to undertake the exclusion of Ulster as a whole from the home rule scheme, how would they explain such extreme cases as Cavan with 74, 000 Catholics out of a total population of 91,000; Donegal with 133,000 out of 168,000 and Monaghan with 53,000 out of 71,000?[45]

Replying for the Opposition, Lansdowne referred to some parliamentary constituencies where the parties were evenly balanced, for example, North Tyrone, and East Tyrone, North Fermanagh and Londonderry City.[46]

The second reading debate extended over four days and the stage was carried by 273 votes to 10. But on the first day in committee, Lansdowne moved a two-clause amendment, providing that the Government of Ireland Act should not apply to the 'excluded area', defined as the whole province of Ulster.[47] This amendment was carried on the same day by 138 votes to 39. A few minor amendments were added on the report stage.

Moving the third reading, Crewe complained that the opposition had never explained how they would meet the objections of the Catholic counties to exclusion.[48] But it was too late: the amending bill, as amended by the Lords, passed its third reading on the same day.

The government was now faced with an acutely embarrassing dilemma. If they presented the amending bill with the Lords' amendments to the Commons on 20 July (the day scheduled for its consideration), the Irish Nationalists would

42 Lyons, *John Dillon*, pp 350–2, gives a succinct account of these developments.   **43** Lyons, *John Dillon*, p. 351.   **44** *House of Lords Debates*, xvi (23 June 1914), 377–382.   **45** Ibid. 381–2.   **46** Ibid., 536.   **47** *H of L. Deb.*, xvi (8 July 1914), 894.   **48** *H of L. Deb.*, xvi (14 July 1914), 1120–1.

surely vote against it, defeat the government and force an election. On the other hand, the Home Rule bill, protected by the Parliament Act, was due to come into law by the end of the parliamentary session. With masterly understatement Asquith's biographer writes: 'A settlement by negotiation had therefore become an urgent necessity for the Government'.[49]

For once Asquith reacted promptly. On 16 July he told the king, who had been pressing for a meeting between Redmond and Carson, that the moment had arrived for a conference between the interested parties to the Irish dispute, and proposed that it be held at Buckingham Palace. The king readily agreed and suggested that Mr Speaker Lowther should preside.

The conference met at Buckingham Palace between 21 and 24 July.[50] The participants were Asquith and Lloyd George for the Liberals, Redmond and Dillon for the Irish Nationalists and Bonar Law, Lansdowne, Carson and Craig for the Conservative and Unionists. Asquith apparently believed he could achieve agreement over the excluded area, but, if so, he was to be disappointed. The conference broke down over 'that most damnable creation of the perverted ingenuity of man – the county of Tyrone'.[51] Redmond pointed out that in Tyrone there were 53,000 Catholics to 40,000 Protestants.[52] Carson in reply claimed that these 40,000 paid three-quarters of the rates. Neither would give way, nor did they accept the Speaker's helpful proposal that they divide the county equally between them. It was clear that the Unionists would not settle for less than permanent exclusion of the six counties.[53]

After the failure of the Buckingham Palace conference Asquith secured the reluctant assurance from Redmond and Dillon that they would try to persuade their party to agree to support the re-introduction of the amending bill (that is, county option) *without* a time-limit.[54] At that meeting Redmond informed Asquith of the unexpected friendliness of the Ulster Unionists at the conference: Craig having warmly shaken Dillon's hand and Carson having tears in his eyes when parting from Redmond. In his account of the day's events in a letter to his young friend, Venetia Stanley, Asquith wrote: 'Aren't they a remarkable people? And the folly of thinking that we can ever understand, let alone govern them!'[55] – a sentiment shared by many British politicians before and since.

Asquith planned to introduce the amending bill in its new form in the Commons on Tuesday 28 July. But on Sunday 26 July occurred the 'Bachelors Walk massacre'. Emulating the UVF, the Irish Volunteers landed guns in broad daylight at Howth, Co. Dublin. In the ensuing clash with the military, three civilians were killed and 38 injured. Nationalist opinion was so outraged that Asquith

49 Jenkins, *Asquith*, p. 318. **50** The fullest account of the proceedings of the conference is to be found in memoranda written by Bonar Law, dated 21–24 July. See B.L.P., 39/4/44. **51** Jenkins, *Asquith*, p. 320. **52** Pease Diary, 22 July 1914. **53** Ibid., 24 July 1914. **54** Jenkins, *Asquith*, p. 321. **55** Asquith to Venetia Stanley, cited in Jenkins, *Asquith*, p. 322.

announced another postponement – until 30 July. However, by that day the war clouds were looming over Europe – the Austrian ultimatum to Serbia had been sent during the previous week – and Bonar Law and Carson requested a meeting with Asquith at which they proposed that the debate on the amending bill be postponed for the time being on account of the international situation. Asquith welcomed the suggestion and Redmond agreed. It was then understood that the home rule bill should become law, but that its operation would be postponed until another amending bill could be passed.[56]

On 3 August Redmond made his famous speech offering the Irish Volunteers for the defence of Ireland ('for this purpose armed Nationalist Catholics in the South will be only too glad to join arms with the armed Protestant Ulstermen in the North'.)[57]

Aware of the widespread British approval for his speech on 3 August, and also of the growing opposition among rank-and-file Liberals to what they regarded as unacceptable appeasement of the Unionists,[58] Redmond pressed Asquith – who was still thinking of further concessions [59] – to get the Home Rule bill on to the statue book. [60] Eventually, Asquith gave way and allowed the bill to receive the royal assent, but coupled with it was a suspensory bill, postponing its operation until a date 'not earlier than the end of the present war', and a pledge that parliament would then make special provision for Ulster.

The two bills passed under the provisions of the parliament Act; the Government of Ireland bill, received the royal assent at the traditional ceremony in the House of Lords on 18 September 1914. The opposition boycotted the ceremony, claiming breach of faith, since Asquith had agreed with Bonar Law that 'controversial' legislation should be deferred until after the war. But some 60 Nationalist, Liberal and Labour members attended, and many more packed the galleries from which arose a deafening cheer, when the clerk of parliament pronounced the ancient formula 'Le roy le veult'. Then 'The excited Commons surged back through the corridor to their own Chamber, and as they did so, someone produced a green Irish flag emblazoned with a golden harp and waved it in triumph above their heads. When they had returned, a Labour member asked in a voice that trembled with emotion, if they might sing "God Save the King", and without waiting for an answer, began to lead the House in the anthem. Parliament was then prorogued; the members filed past the Deputy Speaker and shook hands with him. The time was twenty-five minutes past twelve, and the date 18 September 1914.'[61]

56 Jenkins, *Asquith*, p. 323.   **57** Mansergh, *The Unresolved Question*, p. 82; G.K. Peatling, *British Opinion and Irish Self-Government, 1865–1925: From Unionism to Liberal Commonwealth* (Dublin, 2001), pp 73–81.   **58** David (ed.), *Inside Asquith's Cabinet*, p. 175.   **59** D.G. Boyce (ed.), *The Crisis of British Unionism: Lord Selborne's Domestic Political Papers, 1885–1922* (London, 1987), pp 114–15. Mansergh, *The Unresolved Question*, p. 84.   **60** Stewart, *The Ulster Crisis*, p. 17.   **61** Ibid., p. 231.

We must now consider the oft-repeated question of whether Ireland was on the brink of civil war in the summer of 1914 until the welcome interposition of the Austrian ultimatum turned members' thoughts elsewhere.[62] We must ask to whose advantage civil war would have been? Certainly not the Ulster Unionists', because through the Lansdowne amendment they had already secured the exclusion of the whole of Ulster from the home rule settlement. The Irish Parliamentary Party were equally averse to a military conflict, and had managed to secure effective control of the Volunteers.[63] The UVF were most unlikely to start a conflict without the permission of Carson and the other leaders. The conclusion is that the advent of the First World War saved the British government from an embarrassing situation but did not of itself save Ireland from civil war.

This chapter has told of contrasts between unprecedented vacillation and infirmity of purpose on the part of the British government, compared to steely determination on the part of the Ulster Unionists (though their leader was ambivalent) and almost heroic patience from the Nationalist leaders.

It may therefore be appropriate to end on another contrasting note. After 1917 Cork city and county became a hotbed of republicanism and many melodious rebel songs originated there. But in 1914 and 1915 a song of a very different tenor was sung in the working class areas of the city. It has never appeared in any printed collection.

> Give me down my fourpence, give me down my shawl.
> Give me down my fourpence till I go to the City Hall,
> Right away, right away,
> Right away my jolly old Munsters, right away!
> And when the war is over, what will the slackers do?
> They'll go up to every soldier for the loan of a bob or two.
> Right away, right away,
> Right away my jolly old Munsters, right away![64]

---

62 Jalland, *The Liberals and Ireland*, p. 260. Lyons, *John Dillon*, p. 354.   63 Mansergh, *The Unresolved Question*, p. 85.   64 Personal information to C. O'Leary from the late Ms May O'Driscoll, Willowbrook, Western Road, Cork (*c*.1950). The Munsters were the Munster Fusiliers, the best known regiments in the South of Ireland. The women went to the City Hall to collect mail from the Front, with, of course, remittances from their husbands.

2

1916: the last chance?

Asquith commissioned David Lloyd George, his most nubile negotiator, to make a renewed attempt to persuade nationalists and unionists to agree on an immediate home rule settlement. All the old problems re-emerged. Redmond now conceded the 'temporary' exclusion of six counties, but Ulster Unionists refused to budge on their demand for their permanent exclusion, though they were now reduced definitely to this territorial claim.

> J.J. Lee, *Ireland, 1912–1985* (Cambridge, 1989), p. 37.

This Irish business has been so badly handled. Fancy getting an agreement between Redmond and Carson and then losing it. The agreement itself was a miracle.

> 1st Viscount Esher, *Journals and Letters of Reginald, Viscount Esher*, iv (1916–30) (London, 1938), p. 39, Esher to Lady Esher, 26 July 1916.

Before the Cabinet meeting today, there was a meeting between Carson and the Unionist Ministers. My information is that Carson gave them hell, and gave them notice that he was determined, to stand by the settlement.

> T.P. O'Connor to John Redmond, 21 June 1916,
> Dillon papers, TCD 6741/325.

After the outbreak of war, Ireland and its problems faded from British public consciousness, except for an unseemly row, when Kitchener, the secretary for war, refused to set up an Irish division, although he had approved an Ulster one. So the Easter week rebellion of 23–29 April 1916 took virtually everyone by surprise, not least the three top government officers in Ireland, Lord Wimborne (lord lieutenant), Augustine Birrell (chief secretary) and Sir Matthew Nathan (under-secretary), all of whom resigned forthwith.

The story of the Easter Rising has often been told, Before we discuss the political moves that it provoked, we will make just a few observations.

(1) The efficient cause of the revolt was the split in the Irish Volunteers, following a speech by Redmond on 20 September 1914, in which he appealed to the Volun-

teers to serve 'not only in Ireland itself, but wherever the firing line extends, in defence of right, of freedom and religion'. Thereupon a minority of the Volunteers, led by Eoin MacNeill, seceded and established a separate organization, which continued to call itself the Irish Volunteers (the Redmondite majority were known as the National Volunteers). It was this body that planned the Easter Rising with vague (unfulfilled) hopes of assistance from Germany. But the core group of conspirators, members of a secret republican organization dating from the Fenian rising of 1867, were the real planners of the rebellion and carried it out in spite of efforts by Mac-Neill, as commander of the Volunteers, to stop it. Although the British were to hold Sinn Féin, the separatist group led by Arthur Griffith, responsible for the rising, the leaders of the party were not involved. Lee writes: 'The home rule press and the British succeeded in investing Griffith's moribund Sinn Féin with a degree of authority it had never managed to achieve on its own, by the simple device of branding all rebels Sinn Féiners.'[1] 'Moribund' seems rather harsh. Since its foundation in 1905, Sinn Féin had only contested one by-election (unsuccessfully), but it had secured the election of some councillors, especially in Dublin. After 1910 it split into two groups, the one led by the original leaders, Griffith, Edward Martyn and John Sweetman (who were monarchists), the other a congeries of republican clubs under the influence of Seán MacDiarmada, one of the signatories of the Easter week proclamation.

(2) In the power vacuum that existed after the Rising, Asquith's cabinet proclaimed martial law and appointed as the effectual military governor of the country General Sir John Maxwell, a battlefield warrior with experience in Egypt, the Sudan, and South Africa. It was to prove a disastrous choice. Maxwell established military courts, which sentenced fourteen of the leaders of the Rising to death by firing squad. The executions were spread over a fortnight, which served to alienate public opinion, and the provincial newspapers, which initially denounced the Rising either as a 'German plot' or a 'communistic disturbance', also denounced the severity of the repression.[2] Eventually on 6 May the cabinet ordered the ending of the executions, but the damage had been done.

Maxwell was as obtuse as he was severe. He bombarded the cabinet with reports – on heaven knows what information – of widespread opposition to Redmond and his party,[3] and hinted at a possible renewal of violence, but when questioned at a cabinet meeting on 27 June, he stated that

> with the present military force in Ireland, anything in the nature of a serious rising was impossible, and that a division of good troops in command of an

---

1 Lee, *Ireland, 1912–1985*, p. 38.   2 Lee, *Ireland, 1912–1985*, pp 38–44. See also O.D. Edwards' careful analysis in O.D. Edwards and F. Pyle (eds), *1916: The Easter Rising* (London, 1968), pp 241–71.   3 Reports by Maxwell (24 June, 17 July), Curzon MS, F112/176.

Imperial General could always (apart from foreign invasion and aid) safe-
guard the military situation.[4]

If Redmond had accepted Asquith's offer of a cabinet post in 1915, he would surely have been in a position to influence their deliberations in 1916. This decision was
apparently based on nationalist ideology (like refusing to accept the hospitality of
the viceregal lodge), but as Lee writes:

> There was little logical justification for supporting the war but refusing office.[5]

On 11 May Asquith himself travelled to Dublin. He had already sent over the permanent secretary to the treasury, Sir Robert Chalmers, who begged for a new head
for the Irish administration as soon as possible. Asquith spoke to many officials, Irish
Nationalist members and even the prisoners in Richmond Barracks, and found them
'very good looking fellows with such lovely eyes'.[6] He also travelled to Belfast and
Cork.

On his return to London on 19 May, the prime minister prepared a memorandum
on the Irish issue for presentation to the cabinet.[7] The first part of the memorandum
dealt with the security situation. On the one hand Asquith did not find any 'general
or widespread feeling of bitterness' between the civil population and the military,
while he himself had been received with remarkable warmth by a crowd in Dublin,
and the military and police authorities did not anticipate any further outbreak in the
near future. On the other hand there would be no lasting peace in Ireland until steps
were taken to control and limit the possession of arms.

But that goal could only be attained in the context of a political settlement. Asquith
was impressed by the lord mayor of Belfast's assurance that recruitment for the forces
was at a standstill in Ulster (because of fears of a 'Nationalist invasion of our province')
and wrote that while it was clear that the Home Rule Act, however amended, could
not come into operation until the end of the war, provision had to be made for the
government of Ireland in the meantime. His only substantive recommendation was
that the 'costly and futile anachronism' of the viceroyalty should disappear and be
replaced by annual visits of the royal court to Ireland, and that instead of the dual
authorities, the chief secretary and under-secretary, a single minister be responsible
for the Irish administration.

The cabinet's reaction to this proposal was favourable, and on the next day (22
May) Asquith wrote to Lloyd George, inviting him to 'take up' Ireland, 'at any rate
for a short time', adding that 'there is no one else who could do so much to bring
about a permanent solution'.[8] Lloyd George, then immersed in his duties as minis-

---

4 Asquith to the king, 27 June 1916, CAB 41/37/24. 5 Lee, *Ireland, 1912–1985*, p. 23. See
also Mansergh, *The Unresolved Question*, p. 87. 6 Jenkins, *Asquith*, p. 398. 7 Memorandum by Asquith for cabinet, 19 and 21 May, CAB 37/148/18. 8 Gilbert, *Lloyd George*, p. 320.

ter of munitions, hesitated at first. He refused to go as chief secretary and undertook only to try to solve the problem of governing Ireland. On 22 May he wrote to his brother, saying that he had been asked by the prime minister and Bonar Law

> to take Ireland with full powers to effect a settlement. Rather interesting that when there is a special difficulty, they always pick on me![9]

On 25 May Asquith told the House of Commons[10] that it was of paramount importance that, if it were possible, an agreement 'such as we sought, and sought in vain' before the war should be arrived at between those representing the different interests and parties in Ireland, and that at the unanimous request of his cabinet colleagues Lloyd George had undertaken to devote his time and energies to that end.

Asquith also mentioned that Lloyd George had already got in touch with the Irish parties. On 26 May, at a lunch with Carson and Sir James Craig, he unfolded his plan. The Home Rule Act would be brought into immediate operation in 26 counties. The six counties would remain under the imperial government. That arrangement would remain for the duration of the war. During that period Irish representation at Westminster would remain at 103. The Irish MPs, minus those from the excluded area, would constitute the Irish House of Commons, and there would be a nominated Senate.

On 29 May Lloyd George conveyed the finished draft of his proposals with a covering letter to Carson:

> We must make it clear that at the end of the provisional period Ulster does not, whether she wills it or not, merge in the rest of Ireland.[11]

This sentence has, in the eyes of most Irish scholars, damned Lloyd George as a double-dyed deceiver, since it seems to contradict what he told the Nationalist leaders. But Carson interpreted it as meaning that everything would be considered again after the War.

> The six counties are to be excluded from the Government of Ireland Act and are not to be included unless at some future time the Imperial Parliament pass an Act for that purpose.[12]

In the same week, Lloyd George met four Nationalist leaders, Redmond, Dillon, Devlin and T.P. O'Connor. Apart from Dillon, whose natural pessimism again came to the fore, the other three, especially Redmond, were anxious for a settlement. They agreed to the Lloyd George terms on the understanding that at the end of the war an

---

9 W. George, *My Brother and I* (London, 1958), p. 254.   **10** *Hans* lxxxii (25 May 1916), 2310–12.   **11** I. Colvin, *Life of Lord Carson* (London, 1936), iii, p. 156.   **12** Colvin, *Lord Carson*, iii, p. 167.

imperial conference should produce a permanent settlement. Knowing that any pro-
posal for partition would be difficult to sell to the party, the Nationalists asked Lloyd
George for an assurance that no further concessions would be required from them.
Redmond[13] wrote of the meeting:

> He gave us the most emphatic assurance, saying that he had 'placed his life
> upon the table and would stand or fall by the agreement come to'. He assured
> us also that this was the attitude of the prime minister. We said on that assur-
> ance we would go to Ireland and ask the consent of our people, but not oth-
> erwise.[14]

The southern Irish Unionists were quick off the mark. On 25 May the MPs and peers
from outside Ulster met and decided that they did not like the opening of the

> vexed question of Home Rule, which would involve the disregard of the party
> truce, while the country is engaged in war, and they deprecated any perma-
> nent change being made as a result of recent events without the prior con-
> sent of the Electors of Great Britain and Ireland.

These resolutions were conveyed to Lloyd George in a letter from Lord Midleton,[15] a former Conservative cabinet minister who was the recognized leader of the
southern Unionists. Lloyd George responded quickly and invited Midleton, Lord
Desart and George Stewart, the vice-chairman of the Irish Unionist Alliance, to meet
him on 29 May. He asserted that an Irish settlement had become absolutely necessary, because as a result of Easter Week the mood of Irish-Americans had changed
from 75 per cent siding with the Allies to 100 per cent pro-German, and that the combination of the Irish and German vote could prevent the election in November 1916
of any candidate (like President Wilson) who was pro-Ally.[16]

This argument was later disputed by the Conservative leaders, but was supported by the dispatches of the British ambassador to Washington, who wrote on 19 May:

> The Irish-German alliance will direct its energies to forcing both candidates
> to adopt an anti-British attitude during the term of the election.[17]

---

13 See Gwynn, *John Redmond*, p. 506; Lyons, *John Dillon*, pp 386–8.   14 Gwynn, *John Redmond*, p. 506.   15 Midleton to Lloyd George, 26 May 1916, cited in Kendle, *Walter Long*, p.
99. Midleton had problems nearer home. His unmarried sister, Albinia, became so captivated
by Irish culture that she changed her name to Gobnait Ní Bhruadair and went to live in a small
cottage in Kerry. She survived until 1955.   16 Kendle, op. cit., p. 100.   17 Stephen Gwynn
(ed.), *Letters and Friendships of Cecil Spring-Rice* (Boston, 1929), ii, p. 331. For the tensions
within the American Catholic press see T.J. Rowland, 'The American Catholic Press and the
Easter Rebellion', *Catholic Historical Review*, 81:1 (January 1995), 67–83.

This dispatch was circulated to the cabinet meeting of 29 May by Grey, the foreign secretary. Balfour knew about it,[18] so there was no excuse for other Conservative ministers professing ignorance of it.

Lloyd George told the southern Unionists that this settlement was the price they had to pay to the Empire for the war. Midleton recorded that 'the project did not appear to us at all hopeful',[19] but he did not reject it at this point.

Meanwhile, Walter Long, who had just warned Lloyd George that Carson no longer reflected the views of southern Unionists, met Lloyd George on 30 May. He took some 'rough notes'[20] of the proposals and told Lloyd George that in his opinion no 'Irish Unionists or any English Unionists for that matter' would accept them, but was told that they had been favourably received by both the Nationalists and both sets of Unionists in Ireland.

The cabinet had appointed a small committee to monitor the negotiations, comprising Asquith, Crewe, Lansdowne, Long and Lloyd George. This committee met just once, on 1 May. Lloyd George mentioned the argument about the USA – which Long claimed to have heard for the first time.[21] Lloyd George indicated that Nationalist consent to the proposals was dependent on home rule being brought into operation during the war. Long and Lansdowne repeated their objections with great force.[22]

From the fact that neither Long nor Lansdowne sought a cabinet meeting to discuss the proposals, nor alerted their Conservative cabinet colleagues or Carson to what they perceived to be the inherent dangers, one may reasonably infer that they did not at that stage intend to wreck the scheme. A memorandum by Lansdowne to Asquith on 2 June merely argued that the minister of munitions' scheme would 'provoke a storm of antagonism in many quarters'.[23]

Unaware of these manoeuvres, Carson proceeded to Belfast to lay the scheme before the Ulster Unionist Council on 6 June.[24] He told an unsympathetic audience that the Home Rule Act was then an inescapable reality, and that the 'clean cut' of the six counties was as much as they could hope for. There was much resentment at the abandonment of the Unionists in the three border counties and the South and West, but in the end the Council accepted the scheme. ('We feel, as loyal citizens, that, in this crisis of the Empire's history, it is our duty to make sacrifices.')[25]

But, if Lansdowne and Long had not yet decided on their strategy concerning the Lloyd George scheme, the southern Unionists had by then resolved to 'fight it to the death'.[26] On 6 June seven Irish peers (including Midleton and Desart) and one Union-

18 Shannon, *Balfour and Ireland*, p. 216. Long later claimed that the American issue had never been discussed in cabinet. Memo by Long (23 June 1916), CAB 37/150/15.   **19** Kendle, *Walter Long*, p. 101.   **20** LP WRO, 947/402/6.   **21** Memorandum by Long, 'The Irish difficulty', CAB 37/150/5.   **22** Ibid.   **23** Memorandum by Lansdowne, 2 June 1916, LP WRO 947/402/10; Kendle, *Walter Long*, pp 104–5.   **24** Jackson, *Carson*, p. 54.   **25** Mansergh, *The Unresolved Question*, p. 94.   **26** Letter from Midleton to Selborne, 10 June 1916, cited in D.G. Boyce (ed.), *The Crisis of British Unionism: Lord Selborne's Domestic Political Papers, 1885–1922* (Lon-

ist representing an English constituency (Walter Guinness) signed a resolution in which they viewed 'with the gravest anxiety' the proposed establishment of home rule, because of the 'spread of the rebellion and seditious feeling' and their (gratuitous) assumption that Redmond's authority had been irrevocably weakened. What determined Long was an angry letter from H A Gwynne, the staunchly Unionist editor of the *Morning Post*:[27] 'You used to be my ideal champion of Unionism and Unionist principles. Today you are helping to betray those same Unionists in the South and West of Ireland who have so blindly in the past given you their trust.' Long was deeply hurt by the accusation, which he decried as 'absolutely false' and 'wholly inexcusable'. But Gwynne replied, repeating the charge in effect, 'though not intention'. In the knowledge that others, besides Gwynne, believed that Unionist ministers were cynically betraying Irish Unionists, Long determined to undermine the Lloyd George scheme: 'Over the next seven weeks he worked incessantly to that end.'[28] Long had a broad range of contacts in the Conservative Party and among the southern Unionists.

His chief tactic was to circulate a series of memoranda to Unionist cabinet colleagues and parliamentarians and also to fire off various letters to Asquith and Lloyd George. His main points[29] were that the Lloyd George scheme had not been submitted to the cabinet, that a pledge had been given of no home rule during the war, that the American danger was illusory, that the Irish Nationalists were 'sullen and hostile', and, of course, that the Irish Unionists of the South and West were being betrayed. He studiously ignored Carson's attitude, claiming that he had been 'misled', which did not give much credit to the Ulster leader's perspicacity! Lansdowne[30] for his part asked rhetorically in a memorandum whether this was the moment to concede in principle all that the most extreme Nationalists had been demanding.

Although the bill was not to be published for another month, its main provisions were revealed on 12 June. They included the immediate operation of the Home Rule bill for the 26 counties, the six to be administered by a secretary of state (as would happen 56 years later), Irish representation at Westminster to remain at 103, while immediately after the war a great imperial conference would consider the government of the empire, including the future government of Ireland. The Act would remain in force for twelve months after the war, but the period could be extended.

On 23 June Redmond and Devlin travelled to Belfast to accomplish a far more difficult task than Carson faced: to persuade the delegates to accept the exclusion of the three counties, Cavan, Monaghan and Donegal, from the home rule scheme. This

don, 1987), p. 171.   **27** Letters from Gwynne to Long, 7, 15 June 1916. Long to Gwynne, 11 June. K. Wilson (ed.), *The Rasp of War: The Letters of H.A. Gwynne to the Countess Bathurst* (London, 1988), pp 172–7.   **28** Kendle, *Walter Long*, p. 107; A. Jackson, *Home Rule* (London, 2003).   **29** Memoranda, 16 June 1916, LP WRO 947/402/6; 23 June 1916, CAB 37/150/15. **30** CAB 37/150/11, Lansdowne memorandum, 'The proposed Irish settlement' (21 June 1916).

was achieved through a virtuoso oratorical performance by Devlin and a clear threat by Redmond to resign as leader unless the scheme were approved – as happened by 475 votes to 265. It is interesting to note that while Dublin's leading Unionist newspaper, the *Irish Times*, on that very day (23 June) wrote that Lloyd George had written to Carson, promising that the exclusion of Ulster would be permanent, this had no effect on the delegates (Redmond denounced it as a falsehood). What is even more remarkable is that there is no reference to the famous Carson letter either in the private correspondence of the Nationalist leaders or in their frequent letters to Lloyd George.

From 19 June there was a split among the Unionist members of the cabinet, the ineffectual leader Bonar Law, his predecessor Balfour and F.E. Smith aligning with Carson, whose support for the scheme never wavered, while on the other side were Lords Lansdowne, Selborne, Salisbury, Robert Cecil and Curzon and Walter Long.[31] The spectacle of English aristocrats and landed gentry lining up against the elected leader of the Irish Unionists on a matter concerning the government of Ireland does not appear to have struck any contemporary commentator as remarkable.

The Irish Nationalist leaders were kept aware of developments within the cabinet by letters from T.P. O'Connor. On 20 June O'Connor reported that 'most of the Unionists were working heaven and earth against Home Rule during the War', but that at a recent meeting Carson 'gave them hell'.[32] The differences came to a head at the cabinet meeting on 27 June. Long opposed the scheme on the presumption that the Nationalists did not really accept it; Lansdowne and Cecil on the ground that it was a concession to rebellion. Curzon and Selborne also opposed. On the other side, Bonar Law wondered what was the alternative, and Balfour forcefully argued that the scheme was a Unionist triumph, since it granted them all they had asked for at the Buckingham Palace conference.[33] Asquith then intervened, pointing out that five cabinet resignations in the middle of a war would be 'not only a national catastrophe but a national crime', and suggesting a cabinet committee of himself, Lloyd George, Lord Robert Cecil and the Conservative lawyer, Sir George Cave, to modify the Home Rule bill where necessary. Selborne had already resigned,[34] but the other four dissidents (Long most reluctantly) agreed to await the outcome of the committee. Meanwhile Long noted with amazement: 'Fancy James Craig spending three quarters of an hour here trying to persuade me, with tears in his eyes, to vote for Home Rule.'[35]

The only amendments proposed by the cabinet committee (on 5 July) were special provisions, safeguarding military and naval rights during the War. Asquith and Redmond thought this unnecessary but were prepared to accept it to placate the

31 See Curzon MS F112/176, 21, 24 June 1916.   **32** Gwynn, *Redmond*, pp 509–10.   **33** CAB 37/150/23, Asquith to the king (28 June); Jenkins, *Asquith*, pp 399–401; Shannon, *Balfour and Ireland*, pp 219–22.   **34** Selborne's excuse *inter alia* was that Carson had put far more pressure on his Ulster friends than he was justified in doing, Boyce (ed.), *Selborne diaries*, p. 184. **35** Kendle, *Walter Long*, p. 122.

Unionists. Long and Lansdowne reluctantly withdrew their plans to resign, and Asquith, announcing the agreement to the House on 11 July, thanked them for their patriotism'.[36]

July 10, 1916, was the latest date on which the Redmond-Carson agreement, the only one ever reached between the leaders of the Irish Unionist and Nationalist parties, might have succeeded. Could it ever have succeeded? Mansergh gives five pointers to failure, mainly because several matters, including the crucial ones of territoriality and duration, were not definitively settled. It is hard to believe that with agreement between the two main parties these obstacles would have proved insuperable.[37]

On 11 July, in a deliberately wrecking speech, Lord Lansdowne insisted that 'structural alterations' could still be made to the 1914 Act and claimed that the Defence of the Realm Act with its stringent provisions (which Redmond had promised his followers would be relaxed) should remain in force in Ireland, even after home rule. This, as expected, was regarded as an insult[38] by the Irish Nationalist leaders, all of whose suspicions of government motives were revived.

After all, Lloyd George had previously promised to resign. Then the Unionist leaders regained the initiative, proposing various niggling attempts to delay the bill, and finally (on 19 July) pushing through proposals to make the six-county exclusion permanent and to reduce Irish representation at Westminster to 43.[39]

This was the last straw. On 22 July Redmond wrote angrily to Asquith that these new proposals would be 'vehemently' opposed[40] by the Irish Party. The Lloyd George bill (only published on 21 July) was withdrawn and on 31 July Wimborne was reappointed lord lieutenant and Henry E. Duke, an undistinguished Conservative backbencher, as chief secretary.

The diehards among the southern Unionists and British Conservatives (especially Long) wrecked the Lloyd George scheme. Long's mental processes were difficult to fathom; Jackson sees him as inspired to some extent by sheer love of intrigue as well as his links with the southern Unionists. He apparently sincerely believed that if a home rule parliament were established so soon after a rebellion, it might lead to the break-up of the United Kingdom, but he took no account of Carson's willingness to take that risk. At various stages Long spoke of the Nationalists as either being sullen and unreliable, or as having lost influence in Ireland, but did not appear to recognize that the very policy he was pursuing was the one calculated to make them remain continually frustrated and deprived of influence.

In fact, there was only one Irish by-election between Easter Week and the end of 1916. In West Cork, which two years later was to return a Sinn Féin candidate unopposed, no Sinn Féiner then appeared and the Nationalist won easily (November 1916).

The first extract at the head of this chapter gives the predominant assessment by Irish historians of the Lloyd George negotiations of 1916. But that was not the only

36 Jenkins, *Asquith*, p. 401.   **37** Mansergh, *The Unresolved Question*, p. 95.   **38** Gwynn, *Redmond*, p. 518.   **39** CAB 37/152/1 (19 July 1916).   **40** Gwynn, *Redmond*, pp 520–1.

contemporary verdict. The second quotation is from a perceptive observer on the sideline of politics. One might also quote from Christopher Addison, then a Liberal junior minister, who wrote in his diary on 23 July concerning Asquith:

> He has never displayed firmness or decision in the cabinet on getting the introduction of the Bill decided on.[41]

But perhaps the most poignant comment was made by Frances Stevenson in her diary:

> July 26 1916
>
> The Irish negotiations have fallen through, and D is depressed and worried about the situation. The Irish are angry with him: they think he should have upheld the original terms of the agreement, and I think they have reason to be angry. A large section of people think that D should have resigned when he failed to carry those original terms in the cabinet: he himself told me he would do so if the Unionists refused them.[42]

41 Christopher Addison, *Four and a Half Years* (London, 1934), p. 234. **42** A.J.P. Taylor (ed.), *Lloyd George: A Diary by Frances Stevenson* (London, 1971), p. 68.

1917

The failure of the negotiations of 1916 left a deep impression on the leaders of the Irish Parliamentary Party, 'convinced to their inmost core that they had been tricked and betrayed'[1] by the British government, and especially by the future prime minister. Several months were to elapse before Redmond spoke again to Lloyd George.[2]

On 18 October 1916 Redmond moved what amounted to a vote of censure on the government because of its Irish policy. Though predictably defeated, the motion allowed some Irish members, especially Dillon, to let off steam.[3] The English Conservatives, wrong on so many Irish issues, were wrong in their assumption that the Irish Party had been destroyed by the 'Sinn Féin rebellion'. One influential Conservative boldly asserted that in a future general election Redmond's party would not even win 12 seats out of 103 Irish seats![4] But, although individual members of Sinn Féin, including Arthur Griffith, either participated in or otherwise supported the Rising, its motive force was the clandestine Irish Republican Brotherhood, the organization that masterminded the two rebellions of 1867 and 1916.[5] Also, in the only by-election (outside Ulster) between Easter week and the end of 1916 the victor was a Nationalist candidate, and there was no Sinn Féin contender.[6]

In March 1917 the perceptive Irish correspondent of *The Round Table* remarked: 'There is no such thing at present as a Sinn Féin party. But there are numerous little groups, each a trifle uncertain of the other'. There was Griffith's Sinn Féin, small, disheartened and suspected by several of the other bodies, on account of its pacifist ideas and its willingness to come to terms with non-separatists.[7] The others were the Irish Volunteers (15,000 before the Rising), the Irish Nation League (a splinter group, mainly consisting of west Ulster Nationalists who deserted the party in 1916 on account of Redmond's willingness to consider even the

---

1 Lyons, *John Dillon*, p. 402. 2 Lloyd George was minister for munitions until 9 July 1916, when he succeeded Lord Kitchener (drowned at sea) as secretary for war. 3 *Hans*, lxxxvi (18 October 1916), 675–86. 4 Philip Williamson (ed.), *The Modernisation of Conservative Politics: The Diaries and Letters of William Bridgeman, 1904–1935* (London, 1988), p. 105 (29 June 1916). Bridgeman was a Coalition Whip 1915–16, later Conservative first lord of the admiralty 1924–9. 5 Unfortunately, to date only one historical work has appeared on this remarkable organization, Leon Ó Broin, *Revolutionary Underground: The Story of the Irish Republican Brotherhood 1858–1924* (Dublin, 1976). 6 The results were: D. O'Leary (Nat.) 1866; F.J. Healy (Ind N) 1750; Dr M.B. Slipsey (Ind N) 370. 7 *Round Table*, 7 (March, 1917), p. 374.

temporary exclusion of the six counties from the home rule area) and lastly the tiny Irish Citizen Army.[8]

But all this was to change in the spring of 1917. In February a by-election occurred in North Roscommon, following the death of the parliamentary veteran, J.J. O'Kelly (MP since 1880). Although the Nationalists were very well organized in the constituency, the Sinn Féiners decided to run a separatist candidate, and they chose Count George Noble Plunkett, a former director of the National Museum, three of whose sons had been involved in the Rising, and one, Joseph Mary Plunkett, had been executed. Plunkett may have been an 'elderly, erratic *enfant terrible*'[9], but he was not devoid of political experience, having been an (unsuccessful) Parnellite candidate in the general elections of 1892 and 1895, and a by-election in 1898.[10] Supported by the Irish Volunteers, Sinn Féin and the Nation League, and an enormous sympathy vote, Plunkett won easily, with a majority over both his opponents.[11] He soon announced that he would not take his seat in the House, although some of his followers disagreed as to whether abstention should be a tactical or a permanent policy.

After his overwhelming electoral victory, Plunkett regarded himself as the leader of the separatist movement and summoned a convention to meet in the Mansion House, Dublin, on 19 April, in order to formulate a new national policy for Sinn Féin. Meanwhile, the leaders of the Irish parliamentary Party were only too well aware of the significance of the Roscommon by-election. On 21 February Redmond drafted a remarkable memorandum, pointing out that a separatist policy would lead to 'inevitable anarchy', and the triumph of Ulster unionism, but offering to step down if the Irish people had grown tired of the monotony of being served for 20, 30, 35 or 40 years by the same people. Fortunately Redmond sent the memorandum to his leading colleagues before releasing it to the press; and realizing that such a document could destroy the party, they persuaded him not to publish.[12] The Irish Party had played no part in the manoeuvres that led to the replacement as prime minister of Asquith by Lloyd George on 9 December 1916 which terminated Lansdowne's cabinet career but restored Carson to the cabinet. When Redmond called on the new prime minister, he was disappointed to find him still considering the partition option. Lloyd George was, in fact, much more concerned at the lack of manpower at the front and had not ruled out the introduction of conscription in Ireland.

The Irish question was raised again in the Commons on 7 March 1917, when T.P. O'Connor moved a motion[13] that, pursuant to the Wilson doctrine of the rights of small nations, it was essential immediately to confer upon Ireland 'the

8 M. Laffan, 'The Unification of Sinn Féin in 1917', *Irish Historical Studies*, 17 (1971), p. 356. The entire article (pp 353–79) is an excellent study of a complex subject.   **9** Laffan, op. cit., p. 360.   **10** Laffan does not mention these facts.   **11** Plunkett (SF), 3022; T.J. Devine (N) 1708; J. Tully (Ind. Nat.) 687.   **12** Lyons, *John Dillon*, pp 410–11.   **13** *Hans*, xci (7 Mar. 1917), 425–42.

free institutions long promised' – and tactfully mentioned the 130,000 Irish volunteers in France. Major Willie Redmond seconded in a highly emotional speech, pointing out that 'the great heart of Ireland north and south beats in strong sympathy with France'.[14] But the Ulster Unionist reaction was predictable. What the Nationalists really wanted, claimed Sir John Lonsdale (Unionist MP for Mid-Armagh), was immediate home rule for the whole of Ireland and, if necessary, the coercion of Ulster. The new prime minister, moving a hostile amendment, insisted that Ulster could not be coerced and made the celebrated remark that Ulster was 'as alien in blood, in religious faith, in traditions, in outlook – as alien from the rest of Ireland in this respect as the inhabitants of Fife or Aberdeen'.[15] These remarks and the whole tone of this 'useless, futile and humiliating debate'[16] so annoyed Redmond that he did not wait for a division, but led his followers out of the House.

On 13 April the war cabinet, reinforced by the presence of Carson, Long,[17] the Irish secretary, Duke, and Christopher Addison (a Liberal minister) discussed the Irish problem and decided to set up a small committee, comprising Curzon, Duke and Addison, to draw up a new home rule bill. The committee met several times between 13 April and 4 May,[18] when they agreed on a draft bill, which Duke would present to the cabinet. Some of its provisions were surprising. The six Ulster counties were to be excluded from home rule, and within a specified period after the war, a plebiscite would be held in each county as to whether they wished to be included in the home rule area or not.[19] If excluded areas still remained, another vote would be taken between seven and ten years thereafter, when a simple majority would suffice. A Council of All-Ireland (the first time such an institution was mentioned) would be established on which the 16 parliamentary members for the excluded counties would sit, which would have the power both to recommend the extension of any particular bill passed by the Dublin parliament to an excluded area and the inclusion of all of Ireland in the home rule scheme. The financial powers of the Dublin parliament were to be strengthened by including the power not only to collect taxes, but to retain and use them.[20] High Court judges for the excluded area were to be appointed by the crown.

The war cabinet was to discuss on 13 May the new bill, which *prima facie* seemed more favourable to the Nationalist cause than the Act of 1914. But to the

14 'He poured out his words with the flow and passion of a bird's song', Terence Denman, *A Lonely Grave: The Life and Death of William Redmond* (Dublin, 1995), p. 111. Denman gives the full text of the speech, pp 106–11. **15** *Hans*, xci (7 Mar. 1917), 459. **16** Ibid., 474–82. **17** Long had moved away from his previous position of *immobilisme* in Irish affairs so far as to suggest two Irish parliaments on the lines of Canadian provincial parliaments, with a possibility of a 'closer union'. Long to Curzon, 12 May 1917, Curzon MSS, F, 112/178, ff19–22. **18** Addison, *Four and a Half Years* (London, 1924), ii, pp 340, 367, 380–400. Addison was minister for munitions 1916–17. **19** For a summary of the provisions of the bill see CAB, 24/89, 1–4. **20** Lloyd George said to Scott that 'there would be no difficulty here. The Unionists were not interested in the question of finance'. Wilson, *Scott Diaries*, p. 283.

disgust of at least one member of the cabinet committee, the bill was dropped, and Redmond was offered a convention in Dublin instead.[21]

2

The notion of a convention of Irishmen who 'could work out their own salvation and agree on a recommendation to the government', actually originated with Redmond[22], and was passed on to Lloyd George who immediately made it his own. His letter to Redmond contained two offers, one being the substance of the Curzon committee's bill, the other being a convention. Redmond accepted the latter. Eventually, in a speech to the Commons on 21 May, Lloyd George promised 'a fully representative' convention'.

Redmond pinned his hopes on the success of the Irish convention on this, the last year of his life. He was only too aware of the threat from Sinn Féin. The Mansion House convention, summoned by Plunkett for 19 April, was a somewhat confused assembly.[23] Only 68 out of 277 public bodies circularized were represented, and the delegates listened in amazement to Plunkett's extraordinary proposal for a parallel organization to Sinn Féin, to be called the Liberty Clubs.

Although the Liberty Clubs only caused bewilderment among the separatists and were fused with Sinn Féin at the end of May 1917, that organization, aided by the Easter Week republicans released from English prisons, grew by leaps and bounds. By July there were 11,000 members and by October 250,000.

The next electoral test was in the South Longford by-election on 9 May, following the death of another Nationalist, John Phillips. The Irish Parliamentary Party ran a very strong candidate, Patrick McKenna, a bacon factory owner, very active in organizing the trade. Although McKenna was supported by 'the bishop, the great majority of the priests and the mob [*sic*] and four-fifths of the traders of Longford'[24], he was narrowly beaten by a Sinn Féin candidate, Joseph McGuinness (1,493 votes to 1,461). Needless to say, this result further weakened Redmond's position.[25]

A more serious blow to the Nationalists occurred in another by-election in East Clare (10 July), following the death in action of Major Willie Redmond at Messines. (In one of his few errors Mansergh states that Willie Redmond lost his seat to Eamon de Valera. Willie Redmond kept his seat until his death on 7 June.)[26]

21 Addison, op. cit., ii, pp 380–1. Addison noted pointedly: 'If also the Cabinet meant to run away from the principle of a home rule bill, when they were confronted with one, why did they ever ask us to draft one?' (16 May 1917). 22 S. Gwynn, *John Redmond's Last Years* (London, 1919), p. 260. 23 Laffan, 'The Unification of Sinn Féin', pp 365–8. 24 Dillon to Redmond, 8 May 1917, cited in Lyons, *John Dillon*, p. 415. McKenna was also supported by the Irish-Ireland journal *The Leader:* see Patrick Maume, *D.P. Moran* (Dublin, 1995), pp 37–8. 25 Redmond admitted to a friend that the Longford defeat had been 'a direct challenge to the authority and representative character of the Nationalist Party': Wilson, *Scott Diaries*, p. 289. 26 Mansergh, *The Unresolved Question*, p. 104.

For the vacancy the Sinn Féin party nominated Eamon de Valera, the senior surviving Easter Week commandant, just released from Lewes prison. In the face of this formidable challenge, the central Nationalist Party was paralysed and inactive. The local Nationalists put up a respectable candidate, Patrick Lynch KC, but against de Valera he stood no chance at all and in the polls secured a mere 2,035 votes against de Valera's 5,010.[27] The East Clare electoral disaster was in no way mitigated by an unopposed Nationalist victory in the Dublin South by-election on 6 July, and when, in the following month, another veteran of the Rising, W.T. Cosgrave, easily defeated the Nationalist candidate in Kilkenny City by 772 votes to 392, it was plain 'not so much that the Nationalists were a spent force, as that they were no longer a force'.[28]

After much manoeuvring between the parties, the membership of the Irish convention was fixed at 101, but only 95 attended. The convention assembled at Trinity College on 25 July, met intermittently before a two-month adjournment in October, then again intermittently from January to April 1918. The standard verdict of Irish historians is that it was 'one of the most striking failures in Irish history'.[29] Apart from some notable absentees (Dillon, Carson, Craig and the dissident Nationalists William O'Brien and T.M. Healy), the convention comprized the most representative gathering of Irish elites ever to assemble: 'It confronted, at a high level of civility and intelligence, most of the issues that would baffle later generations of Irishmen, not only in terms of Unionist-Nationalist relations, but in terms of the nature of a Nationalist state.'[30]

Including the 15 government nominees, the convention comprised five from the Irish Parliamentary Party (led by Redmond), five Ulster Unionists (led by H.T. Barrie), five southern Unionists (led by Lord Midleton), four Catholic bishops, of whom Dr Patrick O'Donnell (Raphoe) and Dr Denis Kelly (Ross) were the most prominent, two Protestant archbishops and the Presbyterian moderator, two Irish peers, one each from the chambers of commerce of Dublin, Belfast and Cork, 31 chairmen of county councils,[31] the mayors of six large boroughs, including Dublin and Cork, eight selected by chairmen of urban district councils (including the duke of Abercorn) and 15 nominated by the government, including the president of University College Cork, the provost of Trinity College Dublin, Lords Dunraven and MacDonnell and Sir Horace Plunkett.[32]

---

27 During the campaign 'de Valera issued no election address, but spoke with impassioned vagueness and liturgical solemnity about the proclamation of 1916': T.P. Coogan, *De Valera: Long Fellow, Long Shadow* (London, 1995), p. 94.   28 Mansergh, op. cit., p. 102. Kilkenny City, in 1917, had only 1,702 voters, compared with South Longford's 3,852 and North Roscommon's 7,997 voters, and so should have been more difficult for a new party to capture.   29 R.B. MacDowell, *The Irish Convention, 1917–18* (London, 1970), p. vii. See also J. Turner, *Lloyd George's Secretariat* (Oxford, 1980), pp 83–113.   30 Lee, *Ireland, 1912–1985*, p. 39.   31 Kerry alone did not participate.   32 The last survivor of the convention, the historian Edward MacLysaght, died in 1986, aged 97.

The convention elected as its chairman Sir Horace Plunkett (who had been a Unionist MP between 1892 and 1900, and subsequently vice-president of the Department of Agriculture and Technical Instruction). Now a home ruler, Plunkett determined to be an impartial chairman.

3

Irish historians assert that the convention was doomed by the absence and outright hostility of Sinn Féin, and the intransigence of the Ulster Unionist delegates.[33] However, behind the scenes, members of the cabinet secretariat (the 'Garden suburb') worked hard for its success. The most perceptive was W.G.S. Adams, the first Gladstone Professor of Political Theory and Institutions at Oxford, who had been seconded to the cabinet secretariat for the duration of the war.[34] Adams saw clearly that Sinn Féin was gaining in Ireland at the expense of the Irish Parliamentary Party, and that opinion was hardening among Ulster Unionists. He was convinced that the only remedy would be an all-Ireland parliament with customs reserved to Westminster, at least until after the war. He had many Irish contacts. Through intermediaries Adams approached Sinn Féin, but learned that they would join the convention only if there were a general amnesty for Irish prisoners and a 'really representative' convention. Nothing was to come of this approach, but at least an effort had been made to involve Sinn Féin in the convention.

From the outset Plunkett tried to direct the convention, but soon difficulties emerged. The main sticking points were an all-Ireland parliament and control of customs. The Nationalists wished the Irish parliament to fix and collect all Irish taxation, while Unionists, both North and South, feared the economic consequences of separation from the rest of the United Kingdom. Control of customs was the touchstone. At the same time, the southern Unionists under Midleton had belatedly recognized that their Ulster colleagues were more concerned with the fate of their own province than with that of the rest of Ireland, and were anxious for the convention to succeed.[35] At the beginning of October, at the insistence of the Ulster Unionists, the full convention adjourned and set up a sub-committee, comprising Midleton, three Ulster Unionists and five Nationalists, to discuss the outstanding issues. The sub-committee settled the parliamentary question by a 'craftsmanlike gerrymander',[36] according to which Unionists would have a bare

33 Lee (a not unsympathetic observer) sums up the convention as 'a largely academic exercise': *Ireland, 1912–1985*, p. 39. 34 For Adams' activities in relation to the convention see Turner, *Lloyd George's Secretariat*, chapter 5. Between 1905 and 1910 Adams had worked as a statistician in the Dublin Department of Agriculture and Technical Instruction under Plunkett. 35 After Lord Midleton had independently sought to hammer out a deal with Redmond at the Irish convention at the end of 1917, Carson and the southern Unionists were irrevocably divided': Jackson, *Carson*, p. 57. 36 Turner, op. cit., p. 102; see Duke's memorandum on the convention, CAB 24/47, GT4105 (2 Apr. 1918).

majority in a joint sitting of both Irish houses. But there was no agreement on the issue of 'fiscal autonomy'.

In September the government resumed the policy of arbitrarily re-arresting released prisoners. This led to the death on hunger-strike of Thomas Ashe (another Easter Week commandant), an event calculated to give maximum publicity to Sinn Féin.[37] In October the first Sinn Féin Ard Fheis (party conference) met to devise a new constitution for the movement, and elect the leaders.[38] The conference established the ascendancy of de Valera and the eclipse of Plunkett. Griffith, as expected, stood for the presidency of the party, but was challenged by de Valera and withdrew. Plunkett also withdrew, so de Valera was elected unanimously. In the election for two vice-presidents, Griffith and Fr Michael O'Flanagan won easily,[39] while Plunkett came a poor third. On the constitutional issue a split between monarchists and republicans was avoided through a compromise resolution, proposed by de Valera. This stated that Sinn Féin aimed at a republic, internationally recognized, but once this had been achieved, the people could freely choose their own form of government by means of a referendum.

Immediately after the Ard Fheis the Irish Volunteers were merged with Sinn Féin, and de Valera was elected president. By then Sinn Féin was a well-organized party with a military wing. Tim Healy, 'no bad judge' told Midleton that even on the existing (male household suffrage) franchise Sinn Féin could win 60 per cent of the Nationalist seats and on the new franchise 90 per cent.[40] Although Redmond, in the last six months of his life, bombarded Lloyd George with letters, urging him to counter Ulster Unionist intransigence at the convention – they made no proposals, and tried to block such as were made – he got no satisfaction from the prime minister, who just stalled.[41] After the Irish convention resumed its full sessions in December 1917, essentially two proposals were seriously discussed. The first was a proposal by Midleton on behalf of the former Southern Unionists for an all-Ireland parliament with full control of purely Irish affairs, including taxation, except for 'imperial' matters, including customs, which would be reserved to Westminster. The other was proposed by Lord MacDonnell (a Lib-

---

37 In a letter to Archbishop Davison of Canterbury (11 Oct. 1917), Archbishop John Bernard of Dublin (a member of the convention) asserted that 'it is no exaggeration to say that the number of Sinn Féiners has been increased by tens of thousands', following the death of Ashe, Curzon MSS F 112/178, ff56–8. Bernard claimed that for many southern Unionists 'the question of the Union is as dead as Queen Anne', and they were quite willing to 'join hands' with Redmond (f60).   38 Laffan, 'The Unification of Sinn Féin', pp 377–9.   39 One of the very few Catholic priests to join Sinn Féin. Bishop Michael Fogarty of Killaloe was another.   40 Midleton to Curzon (11 Oct. 1917), Curzon MSS, F112/178, ff42–4. A Speaker's Conference on electoral reform, which sat from October 1916 to January 1917, produced recommendations that were later embodied in the Representation of the People Act 1918 – essentially universal manhood suffrage and votes for women over 30 years old, if they, or their husbands, occupied land or premises of an annual value of £5. This Act almost trebled the Irish electorate.   41 For a summary of the Lloyd George-Redmond correspondence (Nov. 1917 – Mar. 1918) see Mansergh, *The Unresolved Question*, p. 106.

eral ex-under-secretary), which recommended that control of customs and excise should remain with the imperial parliament during the war and for not more than seven years thereafter, during which a royal commission would settle the question. Redmond was so anxious for agreement that he was prepared to accept the MacDonnell proposals, but he was disowned by some leading Nationalists, including Devlin and Bishop O'Donnell, and withdrew in despair. He died on 6 May 1918.

Eventually, the MacDonnell proposals were accepted by a majority in the convention; the minority comprising the Ulster Unionists, and Nationalists supporting 'fiscal autonomy'. Although Lloyd George had written to Plunkett at the end of February 1918, asserting that:

> The government are determined that, so far as is in their power, the labours of the convention shall not be in vain[42]

the final result – a majority report and two minority reports – did not constitute the 'substantial' majority for which the government had hoped.[43]

Before finishing with the Irish convention,[44] it is noteworthy that in the autumn of 1917 a pamphlet by the ex-minister, Lord Selborne, and the indefatigable advocate of federalism, F S Oliver, entitled *A Method of Constitutional Co-Operation: Suggestions for the Better Government of the United Kingdom*, recommended a federal United Kingdom, both on its own merits and as a means of settling Ireland. It secured support from Walter Long and Austen Chamberlain in the cabinet and from Midleton and Carson among Irish Unionists. W.G.S. Adams, however, argued against Oliver and Carson that Ulster Unionist fears were concerned with internal Irish policies, which a federal solution would not affect.[45]

Carson impressed C.P. Scott no less than Addison with his desire for a compromise:

> The object of his policy was to bring North and South together, but it could only be done tentatively.[46]

---

42 Lloyd George to Plunkett (25 Feb. 1918), CAB 24/47, WX 350.   43 *Hans* civ (9 April 1918) 1364.   44 Although the tone of the entries in the diaries of Lord Oranmore (a Southern Unionist member of the convention) is generally sombre, there are some lighter entries. Describing the final votes, he writes that a Nationalist conventioner, Patrick Governey (a mineral water manufacturer) was so drunk from the after-effects of Redmond's funeral that he could not be sobered up – even with his head put under a tap – and had to be sent to an 'Inebriates Home', without casting his vote. John Butler, 'Select Documents XLV: Lord Oranmore's Journal, 1913–27', *Irish Historical Studies*, 29:116 (Nov. 1995), p. 574.   45 See Turner, *Lloyd George's Secretariat*, pp 103, 107, 112–113. Also Boyce (ed.), *The Crisis of British Unionism*, pp 204–12. The federal question was to come before the cabinet in the following year. See below.   46 Wilson, *Scott Diaries* (19–21 Apr. 1921), p. 277. Addison wrote in his diary for 17 July 1917: 'As I have come to know Carson more and more, my old ideas about him have receded. They were founded on ignorance of the man. A passionate Ulsterman, certainly, but true all through, and extraordinarily clear-sighted and courageous

He told Scott in confidence that what he should propose was that the home rule Act should be brought into operation and certain counties, which he would call X, should be excluded with power to vote themselves in – he did not say whether by bare majorities or something more – and that an Ulster Council of the excluded MPs should have the statutory power to deal with all private bills in concert with the Irish parliament, and to apply to the excluded counties any legislation adopted by the Irish parliament for the rest of Ireland.

Carson felt that this arrangement – virtually identical to the recommendations of the Curzon committee – might well, in the mid-term, allay Ulster Unionist fears and reconcile them to an all-Ireland parliament. At this time he had lost the confidence of the southern Unionists, and his influence with the Ulster Unionists was in decline, with the increasing ascendancy of Sir James Craig and Sir John Lonsdale, chairman of the Irish Unionist parliamentary party since 1915, when Carson joined the government as attorney-general.[47]

In March 1918 John Dillon was elected leader in succession to John Redmond by the Irish Parliamentary Party. In May the two top positions in Dublin Castle were re-allocated: Lord Wimborne, viceroy since 1915, was succeeded on 6 May by 'the little Field Marshal', Lord French,[48] and H.E. Duke, widely though not universally regarded as an unsuccessful chief secretary,[49] was replaced by a Liberal back-bencher, Edward Shortt (5 May). However, none of the three was to exercise the same influence in Whitehall or Westminster as their predecessors had enjoyed before the Easter Rising, and, as 1918 wore on, more and more Irish policies were decided by the British cabinet.

and a great deal more willing to make concessions for the sake of peace than many people suspect'. Addison, *Four and a Half Years*, ii, p. 411. **47** Jackson, *Carson*, pp 53–5. Carson became first lord of the admiralty in 1916, but proved an ineffective administrator, and was made minister without portfolio in 1917. He resigned in January 1918. **48** Midleton told his friend Lord Oranmore that he had been offered the lord lieutenancy, but had declined, 'as they would not give him the full powers he wanted': Butler, 'Lord Oranmore's Journal', p. 575. **49** Duke went on to become a lord of appeal with the title Lord Merrivale. See Eunan O'Halpin's appraisal of Duke ('a political lightweight'), 'Historical Revision XX: H.E. Duke and the Irish Administration, 1916–18', *Irish Historical Studies*, 22 (1980–1), pp 362–76.

4

1918

Everybody knows why Conscription has been introduced for Ireland.
There is not a single man on these benches who believes that this measure will be of the slightest military advantage. Ireland has been brought
in because you want to create a series of battalions of English grandfa-
thers.

Joseph Devlin in the House of Commons,
Hansard, cv (15 April 1918), 94

On 12 April 1918, before the report of the Irish Convention could have reached
them, the British cabinet abruptly decided to introduce a new bill raising the maximum age of conscripts from 40 to 51 and including Ireland, which had been
exempted from the first Conscription Act in 28 March 1916.[1] This inept decision, as described by one historian,[2] meant, in the words of another, that the government 'as so often before, came to the rescue of its opponents'.[3]

On 6 April the full cabinet discussed the issue; the prime minister announced
a plan for the simultaneous introduction of bills involving conscription and home
rule. On 9 April Lloyd George asked the Commons for leave to introduce the
Military Service bill. The main reason was the desperate shortage of trained men
on the western front. Lloyd George invited parliament to pass a measure of self-
government for Ireland, pointing out that the majority voting in the convention
did not represent the 'substantial agreement', which the government desired. He
also lauded the 'young men of Ireland' who had flocked to the colours.[4]

The bill was rushed through the Commons – the only major bill guillotined
during the war – and received the royal assent after seven days' debate. It soon
became clear that the Liberal[5] and Labour[6] members of cabinet and parliament
firmly believed that to be acceptable to Irish opinion the Conscription bill should
be linked to the measure of home rule; while the Conservatives, though pas-
sionately in favour of extending conscription to Ireland, were either lukewarm

---

1 CAB, 23/14,WC 376a (28 Mar. 1918).   **2** R.F. Foster, *Modern Ireland, 1600–1972* (London,
1988), p. 490.   **3** Lyons, *Ireland since the Famine*, p. 392.   **4** *Hans*, civ (9 April 1918), 1337–64.
5 See Addison, *Four and a Half Years*, ii, pp 503–4.   **6** See CAB 23/6 WC 309 (11 April 1918),
George Barnes, a Labour member of the war cabinet whose Glasgow constiuency had many
Irish voters, threatened to resign unless a home rule bill was speedily introduced.

about, or hostile to home rule. Supporters of the latter were enraged when the Conservative home secretary, George Cave, moving the second reading, said: 'If you were to link conscription with home rule not only (the) united forces of all Irishmen, but also many Englishmen and Scotsmen would be against and you would carry neither.'[7]

The chief secretary, Henry Duke, was vehemently opposed to the bill and in a letter to the prime minister, asking to be relieved of his office, stated that conscription 'will produce a disaster', but did not publicize his opinion.[8]

The Irish Nationalists were firmly and unanimously opposed to the Military Service bill. On the day of its introduction John Dillon warned that the decision to bring in the bill 'plunges Ireland into bloodshed and confusion',[9] while another parliamentary veteran, William O'Brien, protested against 'this mad and wicked crime you are proposing to-night to perpetrate on Ireland'.[10]

The opposition to the Conscription bill was compounded by the profound distrust which the two Nationalist leaders, Dillon and Devlin, exhibited towards Lloyd George. Indeed, Dillon went so far as to claim in a letter to the editor, C.P. Scott, that 'the whole business has been an elaborate plot by L G to put off the Irish settlement until he has got America so deeply committed to the war that American opinion will not support Ireland'.[11] This seems unfair, since for Lloyd George the main priority was the prosecution of the war, and Irish affairs were of secondary importance. Nevertheless, Dillon persisted in this conviction in spite of Scott's attempts to dissuade him.[12] On 16 April, at the end of the debate on the Military Service bill, Dillon led his entire party out of the House and back to Ireland. Subsequently they resolved to abandon attendance at Westminster altogether.

Meanwhile Dillon's prediction that 'All Ireland as one man will rise against you' was speedily realized – except in Unionist Ulster. On 11 April the Standing Committee of the Irish Catholic bishops issued a statement, warning the government against entering upon 'a policy so disastrous to the public interest and to all order, public and private'.[13] While the bill was still before parliament, the trade unions in Dublin and Belfast organized large meetings in protest.

On 18 April the lord mayor of Dublin, Laurence O'Neill, summoned a conference at the Mansion House to organize a nationwide campaign of protest against the right of England to conscript Irishmen, while they were being denied the right of self-determination. This conference brought people who had never previously sat down together, including John Dillon and Joseph Devlin (Irish

---

7 *Hans* civ (10 Apr. 1918), 1480.   8 O'Halpin, 'Historical Revision XX: H.E. Duke and the Irish Administration, 1916–18', pp 375. The letter was dated 16 April 1918. Duke's resignation was not accepted until May.   9 *Hans* civ (9 Apr. 1918), 1378.   10 Ibid., 1395.   11 Lyons, *John Dillon*, p. 435.   12 Lyons, *John Dillon*, pp 435–7. See also Trevor Wilson, *C.P. Scott's Political Diaries*, pp 341–3.   13 Dorothy Macardle, *The Irish Republic* (Dublin, 1937), p. 251.

Parliamentary Party), T.M. Healy and William O'Brien (independent National-ist), Thomas Johnson and the other William O'Brien (trade unionists), Eamon de Valera and Arthur Griffith (Sinn Féin). This was followed by a national pledge taken at church doors – somewhat resembling the signing of the Ulster Covenant – on 21 April and a general strike on 23 April. As Lyons asserts:

> Not even in his most strenuous endeavours to solve the Irish problem had Lloyd George ever come near producing such uniformity of views amongst so diverse a group of Irishmen as he did by this one action.[14]

Meanwhile, the war cabinet, absorbed in the last great German offensive on the western front, did not appear to be greatly concerned about the reaction in Ireland. On 11 April they decided to set up a committee to draft another home rule bill, which they thought might take the sting out of the conscription plan.[15] The chairman was the ubiquitous Walter Long and among the other members were Curzon, Smuts, Cave, the Liberal professor H.A.L. Fisher and H.A. Duke. The first meeting was held on 15 April. On the following day Long reported to the cabinet that the committee consisted of so many members, all holding impor-tant jobs, that it was not possible to meet as often as he would wish.[16] So the cab-inet authorized Long to draw up the bill himself and just consult his committee on certain points.[17]

As recounted above, Long helped to wreck the agreement between Curzon and Redmond in 1916. By early 1917[18] he had decided that a federal solution was the only way to deal with the Irish problem without endangering the union. A few days before the opening meeting of the committee, Long sent a memo-randum to the prime minister in which he argued that a home rule scheme would be acceptable to both Nationalists and Unionists only if it were part of a federal arrangement. He went on to argue that if the Irish could be persuaded to look at the question from 'the Empire point of view'

> we shall make things easier for the Unionists in the North, and I believe shall attract an immense amount of support from the Nationalists, and strike an absolutely fatal blow at the Sinn Féin movement.[19]

Long's biographer, though generally affected by the *lues Boswelliana*, states that this 'extraordinary' document 'reveals that mixture of hard-headed under-

14 Lyons, *Ireland since the Famine*, p. 394.   **15** CAB, 23/6, WC 389 (11 April 1918).   **16** CAB 23/6, WC 392 (16 April 1918).   **17** There were five Conservatives on the committee (Long, Chamberlain, Curzon, Cave and Duke), three Liberals (Addison, Fisher and the Solic-itor-General Gordon Hewart), one from Labour (Barnes) and an outsider, General Smuts, who was uniquely a member of the war cabinet, though not a member of either House.   **18** Kendle, *Walter Long*, p. 133.   **19** Undated memorandum (*c.*mid April 1918), cited in Kendle, *Walter Long*, p. 150.

standing of Ireland which Long possessed in abundance and the romantic, self-delusory imaginings which all too often sidetracked him'. He goes on: 'It is impossible to understand how he could have thought that "the Empire point of view" would have any hope of prevailing in an Ireland where Sinn Féin was now openly demanding independence'.[20] Kendle might also have doubted whether Long's understanding of Ireland could have been so profound, since he almost completely ignored the Nationalist representatives.

At the second meeting of his committee Long argued that it was no use attempting to placate either Unionists or Nationalists, since no solution that would please one group would be acceptable to the other. He stated that he was 'against presenting any home rule measure which was inconsistent with federalism'.[21]

Although Curzon, Smuts and Hewart were absent from this meeting, no dissenting voice was raised when Austen Chamberlain supported Long, saying that the 'proper test to apply in framing a home rule bill for Ireland was that provided by a federal reconstruction of the United Kingdom'. But there was no consensus on the financial arrangements for Ireland. A sub-committee comprising Long, Chamberlain and Fisher, helped by Duke, W.A.S. Hewins (ex-director of the London School of Economics, a friend of Long) and Lionel Curtis of the Round Table movement discussed the financial issues on 17 and 18 April.

On 18 April Long wrote an uncharacteristically optimistic letter to the prime minister. He began by saying that progress on the financial issues was more rapid than he should have expected. But he went on that 'the further we go, the more clear it becomes' that if they were to start on a federal system for the United Kingdom, the passage of the bill through the Commons would be much easier; it would be far more aceptable to the British population; it would be easier for nationalist Ireland to accept, and (this was always a priority for Long) it would make the position of Ulster 'infinitely easier'.[22]

Long's letter was discussed by the war cabinet on 23 April.[23] The only member to offer wholehearted support was Chamberlain (who argued that it would be impossible to obtain sufficient Unionist support for home rule, except by adopting the federal principle). Lloyd George expressed support for the federal idea, but would not make it a condition for the passage of a home rule bill. The most damaging criticism came from Curzon and Balfour. Curzon pointed out that the cabinet was being asked to agree to a 'novel constitutional procedure' in order to extricate itself from the Irish difficulties. Balfour replied to Long's claim that there was a growing body of opinion in favour of federalism by saying that the subject had received 'a very imperfect consideration'. Federalism might be appropriate in large countries like the United States, Canada and Australia, but there were neither historical nor geographical reasons for

20 Ibid.   **21** Kendle, *Walter Long*, p. 151.   **22** Long to Lloyd George (18 Apr. 1918), CAB 24/49.   **23** CAB 23/6, WC 397 (23 Apr. 1918).

introducing it in a small country like the United Kingdom. England formed by far the largest part of the Kingdom and Balfour thought that an English parliament, co-existing with a federal parliament, would be unworkable. Finally, he believed that the adoption of federalism would plunge the country into acute controversy over consitutional issues at a time when reconstruction problems should be dealt with.[24]

Bonar Law, as usual, 'hovered on the sidelines of both camps', and proposed that the cabinet adjourn this discussion until the draft bill was available. The prime minister pointed out that the Irish situation raised the biggest constitutional issue on which parliament had been divided for thirty or forty years. It would be impossible, he said, to proceed with conscription until home rule had been dealt with. The war cabinet agreed with Bonar Law's proposal.[25]

The difficulties in adopting the federal principle had been clearly expounded at the cabinet meeting on 23 April; but Long obstinately clung to his federalist obsession.[26] He prepared a further memorandum for his committee, which argued not only for the importance of a federal solution for the Irish problem, but also for federalism as a means of relieving the pressure of business on parliament. However, the committee (9 May) were divided on whether to introduce a home rule bill at all during the war, and the meeting was inconclusive.[27]

2

Meanwhile, the truce between the Nationalists and Sinn Féin had broken down. When the aged Samuel Young, MP for East Cavan, died, Sinn Féin rejected Nationalist overtures that parliamentary vacancies should be filled either by the party in possession or by neutral candidates and nominated one of their best-known members, Arthur Griffith, who was now advocating complete independence for Ireland. Dillon prepared to fight a vigorous campaign in support of the Nationalist candidate, but malign fate stepped in once more,[28] with a proclamation by the new viceroy, Lord French, on 18 May, to the effect that the government had learnt that Sinn Féin was engaged in a plot with the Germans; all the Sinn Féin leaders (except Michael Collins and Cathal Brugha) were arrested. Dillon believed that if it were not for the conscription issue and the arrests,

24 In a letter to the British ambassador in Washington (Lord Reading), Balfour promised that the British government 'will stake its very existence on passing a home rule bill' (17 Apr. 1918): Shannon, *Balfour and Ireland*, p. 238. **25** CAB, 23/6, WC397 (23 April 1918). See also CAB 23/6, WC406 (7 May 1918). **26** Kendle, *Walter Long*, p. 156. **27** Ibid., p. 159. **28** Lyons, *John Dillon*, p. 438. In a Commons debate on an Irish members' motion that British policy towards Ireland was inconsistent with the allied fight for 'the rights of small nations', Asquith, referring to the abortive negotiations of 1916, said that both sides appeared at one moment to be 'within at least a measurable distance of agreement', but again 'that malign influence, which has so often presided at critical moments over the fortunes of Irish history, made itself felt, and the attempt failed', *Hans* cix (29 July 1918), 143.

'we had Sinn Féin absolutely beaten'.[29] But although the Sinn Féin candidate, Griffith, was in prison, he still easily won the by-election.[30] The 'German plot' has not been taken seriously by historians. Once again the British government had acted in the way most calculated to rally maximum support for Sinn Féin.[31] Since the cabinet had already decided to intern the Sinn Féin leaders,[32] the 'German plot' provided them with a convenient excuse.

At the same time the physical force element of Sinn Féin, ostensibly the Irish Volunteers, but at its highest level infiltrated by the shadowy, secret society, the IRB,[33] was simultaneously planning to fight the British, initially at least on the conscription issue, and to infiltrate spies into the police and military. It was at this time that Michael Collins, Richard Mulcahy and Piaras Beaslaoi became prominent. All three were members of the IRB.

As early as March 1918, Cathal Brugha, who had been chief of staff of the Volunteers, went to London to draw up in collaboration with other sympathizers contingency plans to assassinate members of the British cabinet, if conscription were imposed on Ireland[34] – a drastic strategy, which had not been tried since the Cato Street conspiracy of 1820.

In April, Walter Long was asked by the prime minister to liaise between the war cabinet and the Irish administration.[35] He visited Dublin at the end of May and returned with a high opinion of the 'new brooms', French and Shortt, but with grim foreboding about the future of the island.

In a memorandum for the war cabinet Long quoted approvingly from an anonymous Irish nationalist:

> The country is now from one end to the other, in a state of high tension owing to its horror of Conscription, and whatever bill you bring in it will be laughed at and torn to pieces by every home ruler in the country, whether moderate or extreme.[36]

Long met his committee for the fourth time on 4 June. (One member, Cave, had resigned, because he felt the committee was 'ploughing the sands'.) Long told them that it was then impossible to proceed with either conscription or home

29 Lyons, *John Dillon*, p. 440.   30 A. Griffith (Sinn Féin), 3795; J.F. O'Hanlon (N), 2581. 31 Lyons, *Ireland since the Famine*, p. 395.   32 CAB 23/6, WC 395 (19 April 1918).   33 Leon O'Broin, *Revolutionary Underground* (Dublin, 1976).   34 See the recently published account by Joe Good, one of the team: Maurice Good (ed.), *Enchanted by Dreams: The Journal of a Revolutionary* (Dingle, 1996).   35 Kendle, *Walter Long*, p. 164. Lloyd George offered the post of chief secretary to Long in February 1918, but Long refused for 'purely domestic reasons'. Ibid., p. 144.   36 CAB 24/53 (1 June 1918), p. 92. William O'Brien and T.M. Healy had been in alliance with English federalists during the pre-war home rule debates.

rule for Ireland. Some members (Barnes, Curzon and Smuts) thought that the government would be embarrassed, if it just abandoned two well-known policies.[37]

Eventually it was decided to seek advice from the war cabinet, and Long was instructed to prepare a report.

Long reported that his committee had come to the conclusion – he did not mention that it was a majority view – that the dual policy was impossible, given the state of Ireland, but he mentioned the adoption of a federal plan as one of the options.[38]

On 19 June[39] the war cabinet considered Long's report, incorporating an emotional appeal for a federal bill. Curzon opened the discussion by pointing out that he had to make a speech in the Lords on the following day, and while he was prepared to say that the dual policy had not been carried out, what answer would he give to the question whether these measures (regarding conscription and home rule) were only temporarily in abeyance, or not? He added that it would probably be argued that the government had departed from the policy previously laid down. A desultory discussion followed. The prime minister stated that the view of the war cabinet had been that if conscription had been put in force and a home rule bill were defeated, the government would resign. When this decision was reached, the cabinet thought that there was sufficient general agreement ('although in certain quarters it might be no more than sullen assent')[40] to carry a home rule measure. Since then, however, two things had happened – the Sinn Féin conspiracy with Germany and the Catholic bishops' challenge to a decision, which, it had always been assumed, rested solely in the sphere of the imperial government.

Other speakers indicated that they were in favour of delaying, but not abandoning the dual policy. The prime minister stated that he would like to wait until he heard the views of an all-party deputation on federalism before deciding whether to consider the question further. The war cabinet, therefore, instructed Curzon to state in the Lords that they should proceed on the assumption that the dual policy had not been abandoned, 'although the government must be the judge as to the time and method of its application'.[41] The cabinet also decided to postpone the consideration of federalism until after Lloyd George had received the deputation.[42]

37 Kendle, *Walter Long*, p. 160.   **38** CAB, 24/54 (14 June 1918).   **39** CAB 23/6, WC 433 (19 June 1918). Also CAB 24/54 (19 June 1918).   **40** CAB 23/6, WC 433 (19 June 1918).   **41** CAB 23/6, WC 433 (19 June 1918).   **42** Addison wrote: 'I have heard a good many incoherent discussions at the cabinet, but never one more incoherent than this: Addison, *Diaries*, ii, p. 545. Curzon's speech in the Lords contained the following: 'In these circumstances it was necessary in both respects, I will not say to abandon the policy – that would be a most unfair description of our position; I will not say to change the front – but it was our duty to recognize the facts of the case as they were before us and adjust our policy to them'. *Hans*, Lords Debates xxx (20 June 1918), 330.

When the prime minister met the deputation, led by Thomas Brassey, his reaction disappointed them. He pointed out that before federalism could be seriously considered there would need to be overwhelming support in the United Kingdom, and especially in England, and – what had apparently escaped Long's attention – general mass support simply did not exist.[43]

Lloyd George's response meant that the federalist cause would not be pursued further in 1918, but although other federalists (e.g. F.S. Oliver) could see the writing on the wall, Long obstinately refused. A motion favouring federalism had been tabled for a two-day Commons debate in July, but the debate was postponed until October, and then until 1919. In a further discussion on 29 July,[44] the war cabinet ignored a proposal by the prime minister to set up an 'authoritative body' to inquire into the subject of subordinate legislatures, and decided to postpone further discussion. Incidentally, at this meeting Lloyd George admitted that he had not even seen the bill which had been drafted by Long's committee, 'as the whole of his time since the 21st March had been taken up with the gigantic task of dealing with the problem created by the German offensive on the Western Front'.[45]

3

Mansergh states erroneously that on 16 April 1918, when the Military Service bill passed through both Houses, the Irish members of the Commons as a body withdrew from Westminster, 'never to return'.[46] In fact, they returned en masse in July, and remained until the prorogation on 21 November.[47] On 29 July Dillon proposed a motion that the British government's Irish policy was inconsistent with the 'great principles' for the vindication of which the Allies were carrying on the war, and had 'greatly alienated and exasperated' the Irish people. He urged the House to accept President Wilson's claim that the 'great objects' of the war were the rule of law and the consent of the governed.[48] Dillon's speech covered much familiar ground. In spite of 700 years of 'unparalleled tyranny', Ireland had never forsaken her claim to be a nation, and he referred contemptuously to the arming in Ulster. The government had decided, against the Nationalists' advice, to introduce the Conscription bill 'on that fatal 10th April'. Concerning the German plot Dillon pointed out that the allegations were not new and that they were unsupported by a scrap or iota of evidence. Lastly, he referrred to the government's offer of land to voluntary recruits as a 'swindle'.

43 Kendle, *Walter Long*, pp 162–3.   **44** CAB 23/7, WC 453 (29 July 1918). The war cabinet had before it two papers, one by Barnes, advocating home rule for Ireland and suggesting that federalism was 'a long way off', and a rejoinder by Long claiming that 'the cause of federalism has progressed'. CAB 24/58 and CAB 24/59 (23–24 July 1918). Long's committee had drafted a federal bill for the United Kingdom. See CAB 24/89 (29 Sept. 1918).   **45** CAB 23/7, WC 453 (29 July 1918).   **46** Mansergh, *The Unresolved Question*, p. 109.   **47** Dillon's last intervention was on 15 November 1918.   **48** *Hans,* cix (29 July 1918), 85–6.

The Irish members were not conciliated by the chief secretary's allegation that they were the real people to blame for the state of Ireland, and Shortt then invited derision by asking them to help to make recruiting a success, 'so as to avoid Conscription'.[49] While the Irish Nationalist MPs made predictable speeches, there were two interesting contributions from ex-ministers: Asquith, that Shortt 'hardly appears to appreciate' that the difficulties he encountered 'have been sensibly aggravated, as some of us predicted they would be', by the adoption of contingent conscription for Ireland.[50] Herbert Samuel (who had been home secretary during Easter week) said:

> I cannot refrain from expressing the opinion that the whole handling of this situation, in respect of Conscription, the Report of the Convention, the introduction of the promised bill, and the failure to fulfil that promise, as *one of the gravest cases of the mishandling of a momentous political situation that this House has ever known.* (our italics)[51]

Almost the last major intervention of the Irish parliamentary Party was the debate on the second reading of the Irish Land (Provision for Soldiers) Bill (1918) on 22 October 1918.[52] The background were two proclamations of the stick-and-carrot type, in May–June by Lord French. On the one hand he authorized the internment of all the leading members of Sinn Féin (except Michael Collins); on the other hand, 'We recognize that men who come forward and fight for their Motherland are entitled to share in all their Motherland can offer. Steps should be taken to ensure as far as possible that land shall be available for men who have fought for their country and the necessary legislative measure is now under consideration.'

This offer was immediately interpreted as applying only to those who responded to Shortt's request for voluntary enlistment.[53]

After many parliamentary questions about the 'necessary legislative measure', the bill was presented to the House on 22 October.[54] The chief secretary did his best to repudiate allegations that the bill was a bribe to encourage voluntary recruits. Shortt estimated that about 3,000 men would benefit at a cost to the State of about £1 million. Dillon, in a devastating reminder, asked whether there could be a 'more insane proposal' than one which would bribe 3,000 recent recruits with £500 a head, while ignoring the 170,000 who had volunteered over the previous four years? Apart from the Irish members, only two ministerialists spoke, Shortt and the Irish attorney-general, A.W. Samuels, a Unionist MP for Trinity College.

---

49 *Hans,* cix (29 July 1918), 108–17.   50 Ibid., 142.   51 Ibid., 192.   52 *Hans,* cx (22 October 1918), 694, 706.   53 See Carson's question, *Hans,* cviii (16 July 1918), 893–894. Shortt suggested that if 50,000 volunteers appeared by the beginning of October 1918, there would be no need for conscription. The actual number was not published, but fell far short of this desideratum. 54 For the debate see *Hans,* cx (22 Oct. 1918), 686–734.

The debate was adjourned and never resumed. On 13 November Bonar Law stated that it would be impossible to carry through the bill in that session and it was withdrawn on 18 November.[55]

Once more the British government, by pursuing an absence of policy in respect of both conscription and home rule – which the war cabinet finally decided to be 'impractical at present', on 21 November[56] – greatly assisted their bitterest enemies in Sinn Féin/Irish Volunteers and discredited the moderate party of John Dillon. The war cabinet never seriously considered forcing conscription on Ireland, but they kept the threat alive until hostilities were clearly coming to an end in October. This attitude enabled Sinn Féin to develop its organisation until at the end of the year, with 112,080 members in 1354 clubs, it had largely supplanted the moribund, Irish Parliamentary Party.[57]

Arrests proved 'merely the signal for the old cycle of intermittent and largely ineffective coercion to begin again'.[58] On 3 July the Volunteers, the Gaelic League and the various Sinn Féin organizations were proclaimed as 'a grave menace', and a further proclamation prohibited public meetings and Gaelic games, unless with police authority.

But although the Sinn Féin leaders were imprisoned, those who escaped continued the struggle against the threat of conscription. When the war was plainly drawing to a close in October, and the conscription threat had receded, Michael Collins and his closest colleagues energetically prepared for the forthcoming general election.[59] At the same time, the last leader of the Irish Parliamentary Party 'was hard at work trying to inject some semblance of enthusiasm into his followers'.[60]

When Edward Shortt was moving the second reading of the abortive Irish Land (Provision for Soldiers) bill, he said to John Dillon: 'I ask the honourable member to remember that he may have to work it (the bill)', to which Dillon replied: 'God forbid!'.[61] The true word was spoken in jest.

As an indication of how popular opinion in Cork was turning against the war and the military generally, we may end by quoting two stanzas of a song, entitled 'Salonika', circulating in 1918.[62] It is an obvious parody of the song quoted at the end of chapter one. The title refers to a long-forgotten British expeditionary force sent to Salonika in 1915 in the vain hope of linking up with Serbia, which was already overrun. The force was refused permission to travel through Greece and remained bottled up at the Southern port until the last stage of the war.

And when the war is over, what will the soldiers do?
They'll be walking around with a leg and a half,

---

55 Ibid., 2686, 3315.   **56** CAB 23/7, WC 505 (21 Nov. 1918).   **57** M. Laffan, 'The Unification of Sinn Féin in 1917', *Irish Historical Studies*, 17 (1971) pp 353–79.   **58** Lyons, *Ireland since the Famine*, p. 396.   **59** Coogan, *Michael Collins*, pp 91–2.   **60** Lyons, *John Dillon*, p. 446.   **61** *Hans*, cx (22 Oct. 1918), 702.

And the slackers will have two,
So right away, so right away,
Right away, Salonika, right away me soldier boy.

For they takes us out to Blarney, they lays us on the grass,
They put us in the family way and leaves us on our ass
So right away, so right away,
Right away, Salonika, right away me soldier boy.

The words come from a CD, *Uncorked*, sung by Jimmy Crowley (Free State Records, 1998) (CD007). In the accomanying booklet, there is no indication that the song is a parody.

5

1919

Excepting the unique election of 1921, the general election of 1918 was the most remarkable in Irish political history. It was marked by the greatest single increase in the number of registered voters. The Representation of the People Act 1918 extended the electorate from virtually male household suffrage to universal suffrage for men over 21 years old and women over 30. It was the first election in which all polling occurred on the same day (14 December). It was the election with the largest number of outgoing members declining to stand. When parliament was dissolved there were 68 members of the Irish Parliamentary Party, seven 'O'Brienites', one Independent Nationalist,[1] two Liberals, sitting for Ulster constituencies,[2] 18 Unionists, seven Sinn Féin (six returned in by-elections and Laurence Ginnell, formerly of the Irish Parliamentary Party, who had sat as an Independent Nationalist since 1910 until he joined Sinn Féin in 1917). F.S.L. Lyons lumps together the Liberals with McKean as 'a handful of Independents'.[3] Of the 103 Irish members of parliament on the eve of the dissolution,[4] no less than 42 failed to stand again. They comprised 31 members of Dillon's party – including Patrick J. Whitty (Louth North) and John L. Esmonde (Tipperary North), both of whom had been returned in by-elections since 1915, all seven 'O'Brienites', including the veterans William O'Brien and Timothy M. Healy, both retiring after nearly 40 years in the Commons, one independent nationalist, and the two Liberals. (Sir James Dougherty's parliamentary career, coming after a long career as a senior civil servant, lasted a mere four years.) At least five Unionists stood down.

The high defection rate among the Nationalists was partly due to years of organisational neglect – in many constituencies there had been no contest for many elections, compared with the enthusiastic recruitment for Sinn Féin, masterminded by Michael Collins and Harry Boland – but was also due to a widespread feeling that the old Nationalist Party was already being supplanted by Sinn Féin. MPs in many constituencies had become detached from their constituents. There were wholesale defections of local parties to Sinn Féin. Several outgoing members firmly promised John Dillon that they would stand again, but backed down at the last minute.[5] Eventually, 25 seats went to Sinn Féin without

---

1 John McKean (Monaghan South).   **2** T.W. Russell (Tyrone North) and Sir James Dougherty (Londonderry City).   **3** F.S.L. Lyons, *Ireland since the Famine*, p. 398.   **4** The Representation of the People Act, 1918, increased the Irish representation to 105, and almost trebled the Irish electorate – from 683,767 (Dec. 1910) to 1,926,274.   **5** F.S.L. Lyons, *John Dillon*, p. 449.

a contest – including a clean sweep in Cork (excluding the city) and Kerry – 11 seats in all.

The contrast between the old and the new was most vividly seen in the two manifestoes. Sinn Féin's 'four points' boldly set out the republican ideal – withdrawal from Westminster, opposition to British rule by 'any and every' means, the establishment of a constituent assembly with supreme power, and finally an appeal to the Peace Conference in Paris to recognize Ireland as an independent nation.[6]

Apart from lauding its record, the source of every reform in Ireland since the 1880s, the Nationalist manifesto was essentially anti-Sinn Féin, deriding its political inexperience and obvious pro-Germanism which would surely prevent it from securing support at the Peace Conference.[7]

It was 'a bitter and ugly election, with no holds barred on either side'.[8] Many allegations of Sinn Féin intimidation of opponents were made.[9] Nationalist speakers also regarded Sinn Féin with contempt.[10]

The result of the general election was a landslide for Sinn Féin. Out of 105 seats that party won 73, the Unionists 26, and the Irish parliamentary Party a mere six seats, five of which were in Ulster (in four cases Sinn Féin abstained). But because four Sinn Féin candidates were elected for more than one seat, the total number of new members was 69. Of the 69 successful Sinn Féin candidates, 34 (including the leaders, de Valera and Arthur Griffith) were in prison, but this did not adversely affect their electoral fortunes. Among those still at liberty was their chief organizer, Michael Collins, who had gone over the full list of candidates 'with a fine-tooth comb', to ensure that only those favouring a 'forward' policy were nominated.[11] The party was sufficiently broadly based to deter the new Irish Labour Party (founded in 1912) from standing. So there was no Labour voice in the first Dáil; in the vivid words of the writer, Peadar O'Donnell, Labour 'confused the prompter's stool with a place on the stage'.[12]

The eclipse of the old Nationalist Party meant that of the brilliant generation of MPs first elected between 1880 and 1885 the only survivor was T.P. O'Con-

---

6 Dorothy Macardle, *The Irish Republic*, pp 223–4.   **7** F.S.L. Lyons, *John Dillon*, p. 446.   **8** F.S.L. Lyons, *Ireland since the Famine*, p. 399.   **9** See A. Phillips, *The Revolution in Ireland, 1916–1923* (London, 1923), pp 152–3.   **10** There is no scholarly work on the 1918 election as a whole, while studies of individual elections are very sparse. See Michael Farry, *Sligo, 1914–1921: A Chronicle of Conflict* (Trim, 1992), pp 144–54, and Oliver Coogan, *Politics and War in Meath, 1913–23* (Maynooth, 1983), pp 79–88. Coogan refers both to attempts to intimidate Nationalist speakers and also to a Nationalist who began his address with 'Fellow Irishmen and fellow idiots', to which the crowd reacted by releasing the brake so that the horse-drawn vehicle, speakers and all, whizzed down the hill! (p. 86).   **11** Tim Pat Coogan, *Michael Collins* (London, 1991) p. 92.   **12** Quoted in B. Farrell, 'Labour and the Irish Political Party System: A Suggested Approach to Analysis', *Economic and Social Review*, 1 (1970), 502ff. See also A. Mitchell, *Labour in Irish Politics, 1890–1930* (Dublin, 1974). Before the election Sinn Féin was prepared to give Labour a few seats, mainly in Dublin, on condition that they would adopt the policy of abstentionism.

nor,[13] who sat for the Scotland division of Liverpool. O'Connor took over the leadership of the small Irish group instead of Devlin, then in ill-health.

Other interesting features of the Irish election were the return of Sir Edward Carson for the new Belfast constituency of Duncairn, instead of his old base of Trinity College, since his support was waning among Southern Unionists;[14] the return of a solitary Unionist for a southern territorial constituency (Sir Maurice Dockrell for Dublin, Rathmines) and the nomination of the imprisoned de Valera for no less than four constituencies.[15]

As has often been mentioned, the Sinn Féin victory in 1918 was not quite as impressive as at first glance it appeared.[16] The total turnout was 69 per cent, and only 47.7 per cent of the votes were cast for Sinn Féin. On the other hand, their candidates were unopposed in 25 constituencies, which, if contested, would have increased their share of the total poll.

The newly-elected Sinn Féin members have been described as 'mostly young and unseasoned'.[17] As might be expected in a new revolutionary movement the average age was about 40 (39.6 per cent), 43 per cent were in the professions and 22 per cent in commerce: 'The picture that emerges of the leadership is that it was young, lower middle class and predominantly urban, that is Dublin-based'.[18] The profile of the Sinn Féin leadership was very similar to that of the Dáil as a whole. Only one deputy (Seán MacEntee) came from Belfast. There was just one woman, Constance Markievicz (universally referred to as 'the Countess'), but she had the distinction of being the first woman elected to serve in the British parliament. Subsequently she was to become the first female cabinet minister anywhere.[19]

Sinn Féin was committed to abstention from Westminster. This obviously meant a separate assembly in Dublin, and between the declaration of the results and the first official meeting of what was to be called the first Dáil Éireann, party members still at liberty prepared for the momentous event.

The Sinn Féin executive invited all 105 elected members to attend the assembly in Dublin on 21 January, but unsurprisingly the invitations were ignored by the Unionist and Nationalist members, so the first meeting of the 'Dáil Éireann',

13 O'Connor was the only Nationalist MP to become a (British) privy councillor, and 'Father' of the House of Commons. He survived until 1929. 14 The two elected members for Trinity were Sir A.W. Samuels KC and Sir Robert Woods, both Unionists. 15 De Valera won easily against Dillonin East Mayo and was returned unopposed in East Clare, but was decisively beaten in Belfast, Falls, by Devlin, and scored a derisory 33 votes in South Down. T.P. Coogan writes misleadingly: 'His only reverse came in a third constituency, West Belfast'. Coogan, *De Valera*, p. 116. 16 Lyons, *Ireland since the Famine*, p. 399. 17 Mitchell, *Revolutionary Government in Ireland: Dáil Éireann, 1919–22* (Dublin, 1995), p. 35. Mitchell analyses the Sinn Féin elite, including only 30 Dáil deputies; the remainder of his elite are leading officials, military leaders and those in supporting activities (pp 34–5). Tom Garvin, *Nationalist Revolutionaries in Ireland, 1858–1928* (Oxford, 1987), pp 49–56, provides an analysis of the background of a more diffuse group of 304. 18 Mitchell, op. cit., p. 35. 19 The second was in Denmark in the 1920s.

as the invitation specified, was attended by the members of one party only, and by a mere 27, since the rest were in prison, or absent for other reasons.[20]

The first meeting lasted just two hours and the proceedings were, almost entirely, in Irish. The Dáil unanimously adopted documents prepared by Sinn Féin committees – the Declaration of Independence with its ringing assertion that 'the elected representatives of the ancient Irish people … ratify the establishment of the Irish Republic'; a provisional constitution, which did not mention the word 'republic', but provided for a parliamentary system, with a cabinet, based on a majority within a single, democratically elected chamber (the Dáil); a Democratic Programme variously attributed to the Labour leader Thomas Johnson. It has been excoriated as 'a sweeping and woolly commitment to social and economic progressivism'.[21] Mitchell notes that socio-economic principles, such as the protection of all children from hunger, lack of clothing or shelter and proper education for all were 'simply in advance of the existing concerns of Irish political life'.[22] This was true not just then but for four subsequent decades.

Another document adopted by the Dáil was 'A Message to the Free Nations of the World', appealing for recognition for the new republic. On the following day the Dáil elected a temporary cabinet, awaiting the release of the imprisoned members. Cathal Brugha was elected prime minister (*Priomh Aire*) and nominated a ministry, comprising Collins, Plunkett, Eoin MacNeill and Richard Mulcahy. The assembly also elected Sean T. O'Kelly as chairman (*Ceann Comhairle*), and three delegates to the Peace Conference (de Valera, Griffith and Plunkett), then adjourned.

On the same day as the inaugural meeting of the first Dáil, what Irish historians call 'the War of Independence', or 'the Anglo-Irish War' began, with the killing of two policemen at Soloheadbeg, Co. Tipperary. The history of this conflict does not concern us here. Suffice to say that the conflict initially merely involved sporadic attacks on police barracks and individual policemen and soldiers – 16 in 1919 – escalating in 1920, and ending in a truce in 1921.[23] Between January 1920 and July 1921 400 policemen and 160 soldiers were killed. The military movement, led by Collins, developed *pari passu* with political activity.

Apart from Sinn Féin organs, Irish national newspapers generally derided the new assembly as 'futile and unreal'[24] (British reaction will be discussed later).

20 Mitchell, *Revolutionary Government in Ireland*, pp 10–18, 346. See also B. Farrell, *The Founding of Dáil Éireann* (Dublin, 1971), pp 51–79. From the beginning members of the new assembly were described as 'deputies' of the Dáil. **21** Foster, *Modern Ireland, 1600–1972*, p. 495. **22** Mitchell, *Revolutionary Government*, p. 16. Collins thought that some provisions, such as 'all right to private property must be subordinated to the public right and welfare' were 'too socialistic': Coogan, *Collins*, p. 105. **23** See two classic studies, C. Townshend, *The British Campaign in Ireland, 1919–21* (Oxford, 1978) and D. Fitzpatrick, *Politics and Irish Life, 1913–1921* (2nd edn, London 1998); also the more recent studies, P. Hart, *The IRA and Its Enemies: Violence and Community in Cork, 1916–1923* (Oxford, 1998) and M. Hopkinson, *The Irish War of Independence* (Dublin, 2003). **24** See list of quotations in Mitchell, *Revolutionary Government*, pp 19–21.

The next important development was the springing of de Valera from Lincoln jail on 6 February. The Sinn Féin leadership, especially Collins, were then taken aback. They had assumed that the president of Sinn Féin would remain in Ireland, but de Valera was unyielding in his insistence that his prime duty was to travel to the US to persuade the Americans to support the revolution. Here de Valera displayed a characteristic which was to recur during his long political career – unwillingness to take advice at variance with his own opinion.[25] But he responded to pressure to stay in Ireland for some time and did not leave for America until June.

The next important development followed the death in prison of Pierce McCann, deputy for East Tipperary on 6 March. The British government, alarmed at the prospect of further prison casualties, released all the remaining prisoners on the same day (6 March), so that at the next meeting of the Dáil on 1 April 52 deputies attended, the largest in the period 1919–21.[26] The meeting elected Eamon de Valera as *Priomh Aire*, which in popular usage was translated 'president of the Dáil'. From that time onward he was referred to as 'President de Valera'.[27] The new president chose his cabinet: Collins, Brugha, Plunkett and MacNeill were reinforced by Griffith, W.T. Cosgrave and Constance Markievicz (three non-cabinet directors were also appointed).

Before adjourning, the Dáil passed a proposal by the president to 'ostracize' the RIC and accepted his report urging a continuation of the anti-British policies of Sinn Féin. The Democratic Programme was left in abeyance.

Sinn Féin set great store by the appeal to the Paris Conference for recognition.[28] Of the first three delegates selected by the Dáil in January two were then imprisoned and all three were refused passports, so Seán T. O'Kelly, Dáil deputy and alderman of Dublin Corporation, was selected as the emissary to Paris. He made little headway in Paris, even after he was reinforced in May by George Gavan Duffy, son of the Young Irelander, who had been born and educated in France. The mission was a failure; the new regime was not admitted to the League of Nations and the report of the Paris Conference made no mention of Ireland. In spite of earlier optimism, it ought to have been clear to Sinn Féin that the delegates at the Conference were most unlikely to take any action that would antagonize Britain, and this was particularly true of President Wilson.

It might be appropriate at this stage to advert to the ignorance of foreign (apart from British) politics on the part of Irish politicians generally. (Even in the days

25 Collins reported that 'you know what it is to try to argue with Dev': Coogan, *De Valera*, p. 129. See also ibid., pp 130–4. **26** The sessions on 1–4 April were private. The next public session was held on 10 April. All took place in the Mansion House, Dublin. **27** During his American visit, later in 1919, de Valera was generally designated as 'President of the Irish Republic': Coogan, *De Valera*, p. 132. **28** Mitchell, *Revolutionary Government*, pp 35–41. Lee, *Ireland, 1912–1985*, writes of an earlier Sinn Féin 'Nationalist Letter to President Wilson': 'The case Sinn Féin presented to him in this letter would have been an insult to the intelligence of a lesser mind than that of a president of Princeton University' (p. 41).

of the Irish parliamentary Party John Dillon was alone among the leaders with both a knowledge of and interest in foreign affairs.) The reasons are not far to seek. In both Great Britain and Ireland the academic study of politics was then in its infancy. The first holder of a chair in Politics at an English university was Professor W.G.S. Adams,[29] who as a member of Lloyd George's 'Garden Suburb' played an important role in Anglo-Irish relations between 1916 and 1918. (It was not until 1948 that a lecturer in Political Science was appointed at a university in Dublin, Trinity College.)[30] Moreover, serious students in Ireland did not have access to comparative works on political systems, the first of which in English was Herman Finer's *The Governments of the Greater European Powers*, published in 1931.

The Irish newspapers too were devoid of serious analysis of European or American politics. They took the reports from the press agencies, but that was all. As late as 1954 the then German ambassador to Ireland informed his superiors that Irish newspapers lacked a perspective on world affairs, mainly because of 'the inadequacy of Irish journalists, who are intellectually incapable of comprehending world events'. He went on: 'The papers have no foreign correspondents … The journalists are badly paid, so that they lack the incentive to improve their knowledge, and they have no resources to undertake foreign travel.'[31]

This situation, which in the eyes of a perceptive foreign observer, reduced the Irish press in general to the level of a provincial press in 1954, *a fortiori* existed in previous decades. An improvement did not occur until the 1960s when the national newspapers employed foreign correspondents and first produced analytical articles on foreign matters.

Following the second Dáil session in April there were seven more meetings in 1919, all but one of which were private. Meanwhile, the new government established its departments in various locations in Dublin and prepared to launch a public loan of £250,000[32] organized by Michael Collins (September 1919). It was oversubscribed by over £120,000.[33] In the same month the British government adopted their first repressive measure against the new regime by proscribing the Dáil.

In June 1919 there emerged the only significant political group in Ireland to advance a different political solution to Sinn Féin, the Unionists and the now

---

29 Adams held the Gladstone Chair of Political Theory and Institutions at Oxford (1912–33), and was warden of All Souls (1933–45). Earlier (1905–10), he was superintendent of statistics and intelligence in the Dublin Department of Agriculture. He died in Co. Donegal in 1966. **30** Dr Basil Chubb, an Oxford graduate. He was promoted to professor in 1961. See also C. O'Leary, 'The Teaching of Politics in Ireland': (iii), Queen's University, Belfast', *Irish Political Studies*, 4 (Dublin, 1986), p **31** Lee, *Ireland, 1912–1985*, p. 607. **32** This loan was separate from the $5 million raised by de Valera in the US. See below, p. 85. **33** See Mitchell, *Revolutionary Government*, pp 57–65. An extra £130,000 came from American subscribers. To see this in context one should remember that to raise £10,000 was regarded as 'a big achievement' for the Irish parliamentary Party. See A. McCarthy, 'Michael Collins – Minister for Finance 1919–22', in G. Doherty and D. Keogh (eds), *Michael Collins and the Making of the Irish State* (Cork, 1998), pp 52–67.

defeated, Irish Parliamentary Party. The Irish Dominion League was founded by Sir Horace Plunkett and introduced to the people in a new journal, the *Irish Statesman*[34] edited by George Russell (Æ) and sponsored by Plunkett, in its first issue (28 June 1919). The manifesto of the League stated its aim 'to promote the immediate establishment of self-government for Ireland within the Empire', as the only possible remedy for 'the present disorder and unrest'. The status of a self-governing dominion for the whole of Ireland would both recognize the distinctive Irish nationality and offer it a place in the Commonwealth. The manifesto listed the economic disadvantages of a republic; apart from its impracticality, it would expose Ireland to a tariff war with England. On the other hand, a dominion would control all legislation for Ireland, as well as levying and collecting all taxes and customs duties. Ireland would cease to be represented at Westminster, which would still control defence. Ulster separatism would be conciliated by adequate safeguards for minorities in a single Irish parliament. But if Ulster Unionists proved obdurate, public opinion in Britain and Ireland would no longer sanction their permanent veto.

Signatories[35] included peers (Lords Ashbourne, Fingall, Monteagle), ex-MPs (Plunkett, Stephen Gwynn, Sir Thomas Esmonde), prominent Unionists (Sir Hutcheson Poe, Sir Algernon Coote), also Professor Mary Hayden, the surgeon Sir Thomas Myles, Mrs Mary Kettle, T.W. Westropp Bennett, later a Fine Gael member and *Cathaoirleach* (chairman) of the Senate of the Irish Free State, and A.M. Sullivan, the second-last serjeant at the Irish Bar.

The editorial in the *Irish Statesman* discerned in the manifesto of the dominion League 'a new portent and a new hope'.[36] But difficulties soon emerged. On 30 June Sir Hutcheson Poe resigned from the League in a letter addressed to Plunkett. He admitted that up to a week previously he was fully prepared to support the League, but the murder of an RIC inspector convinced him that the moment was 'inopportune' for a new contentious movement.[37] While a merger with Stephen Gwynn's tiny Irish Centre Party[38] was encouraging, Sinn Féin had already identified the main weakness of the manifesto: that unlike other dominions Ireland would not have control of defence, since this would be 'impractical'.[39]

As the summer months passed, the *Irish Statesman* grew impatient with the government's 'contemptuous irresponsibility' in refusing to declare a policy for Ireland.[40] When at last a cabinet committee was appointed to make recommendations for the future government of Ireland[41] the paper lamented that several

---

34 *Irish Statesman*, 1:1 (28 June 1919). **35** See W. Wells, *Irish Indiscretions* (Dublin, 1923) for an appraisal of the signatories as indicating the remarkable drift of Irish opinion towards the Left (p. 92). **36** *Irish Statesman*, 28 June 1919. **37** *Irish Statesman*, 5 July 1919. **38** The Irish Centre Party was founded by the ex-MPs, Stephen Gwynn, Hugh Law and Sir Walter Nugent. After this date its fortunes merge with those of the Irish Dominion League. See Wells, op. cit.: pp 89–91. **39** *Irish Statesman*, 5 July 1919. **40** *Irish Statesman*, 2 August, 16 August 1919. **41** See below, pp 86–90.

members had no personal acquaintance with Irish affairs.[42] On 29 October Plunkett gave a speech to the National Liberal Club in London, arguing for full dominion status (apart from defence) for the whole of Ireland.[43] But it soon became clear that the British government then had no intention of creating an Irish dominion.

In spite of contributions from Yeats, Shaw[44] and other writers of quality, the *Irish Statesman* folded up at the same time as the party it supported, in 1920.

The general election of 2 December 1918[45] produced a parliamentary upheaval in Great Britain, as in Ireland. Before the polling the Conservative Party and the Liberals supporting the Coalition agreed to fight the election on a 'coupon' (a common manifesto), while they were opposed by the minority of Liberals (under the leadership of Asquith), and the Labour Party, not to speak of the Irish representatives. The result was a landslide for the 'coupon', which secured the election of 478 MPs out of 707. The Asquithian Liberals were decimated; only 28 candidates were elected, while 225, including Asquith and several former cabinet ministers, were defeated. The Labour opposition elected 63 members, while Conservatives, opposed to the Coalition, returned 23.[46] Lloyd George was still prime minister of a coalition government, but his parliamentary group was reduced to 133 'Coalition Liberals', compared to 335 'Coalition Unionists'.[47] (In the election of December 1910 Conservatives and Liberals both secured 272 members in the Commons.) The large increase in the Conservative ranks meant that the prime minister was to a greater extent than previously a prisoner of the archetypal imperialist party.[48] This imbalance is rarely appreciated by Irish historians.

The new government was formed on 10 January 1919.[49] The Conservatives had ten ministers, including the holders of the crucial offices of the treasury (Austen Chamberlain) and the foreign office (Curzon); but the Coalition Liberals had nine, including Churchill (War and Air), Addison (Local government) and E.S. Montagu (India), while the Irish chief secretary, Shortt, was promoted

42 *Irish Statesman*, 18 October 1919. **43** The *Irish Statesman* carried the full text of the speech (8 Nov. 1919). **44** George Bernard Shaw wrote an article for the *Statesman* in which he argued for a dominion rather than devolution in Ireland ('You cannot make either a country or an individual free by making out a list of the things they are to be allowed to do'). He supported the constitutional arrangements of the United States or Australia, where the states, not the federal governments, enjoyed residual powers. A later correspondent so confused the issue that Shaw wrote another letter in his idiosyncratic style, urging the correspondent to 'go to bed and stay there until the Irish question is settled': *Irish Statesman*, 25 October, 8 November, 22 November 1919. **45** Figures for the 1918 election are taken from David and Gareth Butler, *British Political Facts, 1900–1994* (London, 1994), p. 214. They differ from those in Mansergh, *The Unresolved Question*, p. 118. **46** There were also ten successful 'Coalition Labour' candidates. **47** These numbers exclude the 25 Irish Unionists. **48** On 8 July 1920 a significant minority of Conservatives opposed a Commons motion upholding the dismissal of General Dyer over the massacre at Amritsar. Carson led the opposition. The *Irish Independent* was to present the activities of the Black and Tans as proof that the government was pursuing an 'Amritsar policy' in Ireland. **49** Lloyd George and Bonar Law agreed that until the Peace Conference at Paris was concluded – the British delegates included Lloyd George, Bonar Law and Balfour – to continue the war cabinet (Lloyd George, Law, Curzon, Austen Chamberlain and Sir Eric Geddes). Normal cabinet government was resumed on 31 October 1919.

to home secretary and was succeeded by Ian Macpherson, a Liberal junior minister at the War Office.[50] The coalition election manifesto laid down two impediments to a future Irish settlement – a refusal to contemplate an Irish state outside the British empire or the forcible submission of the six counties of Ulster to a home rule parliament. Otherwise 'all practical paths' would be explored.[51]

After the election English newspapers failed to appreciate the significance of the Sinn Féin victory and the virtual elimination of the Irish parliamentary Party and concentrated instead on the technical problem of what action parliament could take against the unprecedented number of professed abstentionist members. (The answer was nothing, since any declaration of invalidity followed by a fresh election would undoubtedly result in another Sinn Féin victory.) The first meeting of the Dáil was ridiculed as a farce by all leading London newspapers[52] except the Asquithian *Daily News* and *Westminster Gazette*.

At the opening of the new parliament the prime minister's speech made no reference to Ireland.[53] In the same debate Devlin, who took over the leadership of the truncated band of Nationalist MPs, denounced the government for having destroyed the possibility of an Irish solution.[54] Further attempts by Devlin to elicit an answer as to the new Irish policy were fruitless. It appeared that the attentions of British politicians early in 1919 were absorbed by the Peace Conference in Paris.

However, the government did introduce at least one constitutional innovation for Ireland – the adoption of proportional representation for the local government elections, due in the following year. The Local Government (Ireland) bill was introduced by the attorney-general for Ireland, A.W. Samuels KC (Unionist member for Trinity College), on 24 March.[55]

The single transferable vote had been prescribed for elections in constituencies, returning more than two members in the Irish House of Commons and for the entire Senate by the Home Rule Act of 1914. Samuels made no reference to that provision, but stressed that the various suspensory acts would soon expire, and so that the new local government elections would produce 'the best possible form of local government' and 'in view of the dissatisfaction which largely prevails in the country', the government had decided to introduce PR for these elections. He added that since the success of the Sligo experiment[56] such a proposal would be universally acceptable to the whole Irish press.[57]

---

50 In the cabinet George Barnes was the solitary Labour minister.   51 Cited in Mansergh, *The Unresolved Question*, p. 120.   52 See Mitchell, *Revolutionary Government*, pp 20, 23–4.   53 *Hans*, cxii (11 Feb. 1919), 67–81.   54 *Hans*, cxii (12 Feb. 1919), 145–6.   55 *Hans*, cxiv (24 Mar. 1919), 99–103.   56 By a private bill passed in June 1918, STV was prescribed for future elections in the borough of Sligo, partly to encourage more participation by the Protestant minority. At the borough elections in January 1919 the Ratepayers' Association, not Sinn Féin, headed the poll. See C. O'Leary, *Irish Elections, 1918–1977* (Dublin, 1979), pp 7–8.   57 Although the daily papers, the *Freeman's Journal*, the *Irish Independent* and the *Irish Times* were in favour of the change, it

The only opposition to the bill came for the Ulster Unionists, who moved a hostile amendment.[58] PR was 'un-English', and had been excluded from the representation of the People Act 1918.[59] Thomas Moles predicted (correctly) that Sinn Féin would sweep the country in local elections as in parliamentary elections, while Carson professed himself unable to understand the system. The Nationalist remnant supported the bill, Captain Redmond asserting that the Unionists were afraid that Derry City and Tyrone would return Nationalist councils, while Devlin said that if the poor people of Sligo could understand PR (only 40 spoiled votes), why could not Carson?

Sir Maurice Dockrell, the newly elected Unionist member for Rathmines, pointedly distanced himself from the Ulster Unionists by supporting the bill as giving representation where it did not exist. Replying, Samuels admitted the purpose was to blunt the edge of Sinn Féin success.

The amendment to omit the reference to PR was supported only by the Ulster Unionists and six English Conservatives. It was opposed by the pro-government members, the Irish Nationalists, the Labour opposition, two southern Unionists (Dockrell and a member for Trinity College, Sir Robert Woods) and in a thin House was defeated by 170 votes to 27. The second reading was then carried without a division and the bill speedily passed into law.

3

During most of 1919 few revolutionary activities occurred in Ireland. There was only one (formal) public session of the Dáil,[60] the new departments were largely clandestine, and the machinery of the 'counter-state' developed slowly until the return of the president from the United States at the end of 1920, and the availability of large funds from the American loan and other loans in Ireland.[61]

Various attempts to secure a statement on government policy on Ireland in the House of Commons were unavailing in the first half of 1919. On one occasion the chief secretary asserted that no steps could safely be taken to alter the system of government there, so long as the country remained in its existing state.[62] Later he insisted that Ireland was 'a contented country' – except of course for the revolutionaries.[63] The only decisions of importance taken during these months were to issue proclamations attempting to suppress Sinn Féin and proclaiming martial law in County Tipperary – both opposed by H.A.L. Fisher, president of the Board of Education.[64]

was denounced as an attempt to weaken 'the voice of the nation by weakening the majority, which is the expression of that voice' by Chanel (Arthur Clery), *The Leader* (1 Mar. 1919). **58** For the debate see *Hans,* cxiv (24 Mar. 1919), 99–183. **59** Except for elections to university seats – until their abolition for Westminster elections in 1950. **60** To welcome a visiting delegation of American politicians. **61** De Valera's American loan raised just over $5 million. Coogan, *De Valera*, p. 159. **62** *Hans,* cxv (3 Apr. 1919), 1541. **63** *Hans,* cxv (15 May 1919), 1730–1. **64** CAB, 23/12, WC 567a (14 May 1919); WC 585a (26 June 1919).

The cabinet's attention was again focused on Ireland on 5 August.[65] To George Barnes' suggestion (supported by Fisher) that a committee be immediately set up to consider the Irish question, since the problem could not await a settlement of the general question of devolution, Lloyd George grumbled that it would be very difficult to produce a policy before the recess, and the war cabinet had its hands full with the issues of profiteering, finance and housing. With unusual optimism Churchill advised delay, since the Irish people were not only very prosperous, but were beginning to lose faith in Sinn Féin!

The issue came up again at the cabinet meeting on 25 September.[66] On being informed that the Government of Ireland Act 1914 would automatically come into force with the ratification of the last peace treaty, the cabinet agreed that while it could not be simply repealed the Act 'was not acceptable to any of the interests concerned', and the matter must be dealt with in the following session, owing to the dangerous situation in Ireland. The secretary to the cabinet was instructed to circulate the draft bills of 1917 and 1918.

On 7 October 1919 the matter was discussed by the small War cabinet[67] at one of its last meetings.[68] It was realized that some declaration of Irish policy by the government would be expected early in the next session. Two points of view emerged; one that then conditions were 'wholly unfavourable' to any settlement in Ireland, and since any new policy would be bound to fail, it would be better to admit frankly that the time was not ripe for a settlement and merely introduce a law further postponing the operation of the 1914 Act. On the other hand, it was argued that 'there would be no recrudescence of steady and sane opinion' in Ireland until a home rule Act was passed. This view prevailed and a cabinet committee of ten members was set up, again under Walter Long's chairmanship, with a wide remit to examine all options for Irish policy.

Since this committee was the one that drafted the government of Ireland bill, which partitioned Ireland, it is worth careful examination. Although the Conservatives were in a clear majority in the Commons, the committee was evenly divided, with five Conservatives,[69] four Liberals[70] and one Coalitionist Labour.[71] The lord lieutenant (French) and the chief secretary (Macpherson) were members *ex officio*.

When the committee started its deliberations, it had before it a number of documents, including the draft bill of 1918 for a federal system. This clearly

---

65 CAB 23/12, WC 606a (5 Aug. 1919).   **66** CAB, 23/12, WC 624 (25 Sept. 1919).   **67** The war cabinet was disbanded on 31 October 1919. At that time its members were Lloyd George, Curzon, Bonar Law, Austen Chamberlain, and Sir Eric Geddes – all Conservatives except Lloyd George. The war cabinet (of five to seven members) had been set up on 6 December 1916, when Lloyd George became prime minister.   **68** CAB, 23/12, WC 627 (7 Oct. 1919).   **69** Walter Long, Lord Birkenhead, Sir Auckland Geddes, Sir Robert Horne, and Sir Laming Worthington-Evans – all members of the cabinet.   **70** H.A.L. Fisher, Edward Shortt and Sir Gordon Hewart (ministers), and Frederick Kellaway (backbencher).   **71** George Roberts (former minister of labour).

was inspired by the chairman. But the system proposed resembled the system of devolution to be imposed on Ireland in 1920, rather than a truly federal system.[72] The 'local national parliaments' were precluded from legislating on the crown, peace and war, defence, treaties, titles of honour, treason, foreign trade, postal services, coinage, income tax, customs and excise, or to pass any law interfering with religious equality. Furthermore, the lord lieutenant in Ireland (or the high commissioner for the other three countries) was to appoint the executive council and could veto any legislation passed by the local parliament, while the parliament at Westminster was to remain supreme, even in respect of local services.[73]

In a true federation, such as the United States, the central legislature could not override the state legislatures, where the powers of the states were concerned and the state governors did not always have the veto power on legislation. (Also, the powers not granted to Congress, nor prohibited to the states resided in the states, or the people.)

Long was clearly against granting dominion status to Ireland, as some Liberals desired, but what he regarded as federalism seems more properly to be described as devolution.[74]

At its first meeting the committee decided that the mere repeal of the 1914 Act was not an option, but some alternative policy must be suggested.[75] After eight meetings they sent their first report to the cabinet.[76] The two imperatives were the continuing unity of the empire and the absence of coercion for Ulster. Accordingly, they dismissed the possibility of an all-Ireland parliament, even with guarantees for Ulster on the basis of previous bills of 1886, 1893 and 1914, because of consistent opposition from Ulster. They also dismissed the possibility of county option for Ulster on the ground that the area to be excluded would 'almost certainly be administratively unworkable' – although the sovereign state of Luxembourg was smaller in area and population than the four Protestant counties. The third possibility of two parliaments, one for the three southern provinces and one for the province of Ulster, was recommended together with a Council of Ireland (as in the Curzon committee's proposal), composed of members of both parliaments, to discharge certain functions, 'and mainly to promote as rapidly as possible and without further reference to the Imperial parliament the union of the whole of Ireland under a single legislature'.[77]

The belief of the committee in the eventual reunification of Ireland is spelt out over and over again in the report. Their proposal would 'enormously min-

72 CAB 24/89, G.T. 8239 (29 Sept. 1919).   **73** CAB 24/89, G.T. 8239 (29 Sept. 1919).   **74** Kendle writes: 'It was clear [in mid-1919] that Long was not advocating pure federalism but was interested only in devolution with a retention of sovereign powers at Westminster': *Walter Long*, p. 175. **75** See summary of first meeting: LGP, F/24/1/19 (22 Oct. 1919).   **76** See CAB 27/68 for committee meetings. The first report is to be found in CAB 27/68, C.P.56 (4 Nov. 1919).   **77** CAB 27/68, C.P.56 (4 Nov. 1919).

imise' partition, which would not occur if any part of Ireland were under British rule. The committee attached 'the greatest importance' to doing everything possible to promote Irish unity, an object not only desirable in itself, but essential for the scheme to win the support of moderate nationalists.

To this end the committee made two further proposals – that for one year only certain services, which it would be specially undesirable to divide (agriculture, technical education, transport, old age pensions, health and unemployment insurance and labour exchanges) should be reserved to the imperial parliament, but the ministers concerned should consult the Council of Ireland. The Council would have the power to deal with private bill legislation. They also proposed that even in the first year the two parliaments could pass identical legislation to transfer any of the reserved services to the Council of Ireland, and that the two parliaments, acting together, could change the constitution of the Council to make it an elective body.

Finally, the committee followed previous Acts in exempting from Irish control the crown, peace and war, defence, treaties, honours, treason, foreign trade, coinage, wireless and the external postal service. Other services reserved at the outset were income tax, customs and excise and the internal postal service.

These last powers might be transferred to a single Irish legislature, if one should be set up. The committee also recommended that Irish representation at Westminster be continued. But further details, especially on finance, would be examined later.

The committee considered itself to follow the principles of self-determination, as outlined at the Versailles Conference. They were providing immediately states rights for the two areas with a link between them, and also the power to achieve Irish unity on any basis from federalism to qualified dominion status (ie without the defence power).[78]

Mansergh provides a detailed account of this committee meeting.[79] He reproduces the extract which they included from the Coalition election manifesto of November 1918, but does not mention the fact that while the manifesto laid down the condition accepted by all English political leaders, that the six counties of Ulster would not be coerced into submission to a Dublin parliament, the committee recommended a separate parliament for the nine counties. John McColgan also deals with this report and appears to doubt the sincerity of the 'contrived' appearance of promoting Irish unity.[80] But the Long committee had no need to dissemble. Cabinet communications were confidential, and at that time were kept from the public for 70 years. Nor is their sincerity questioned in private correspondence.

---

78 See CAB 27/68, C.P.56 (4 Nov. 1919).   79 Mansergh, *The Unresolved Question*, pp 120, 123–5.
80 J. McColgan, *British Policy and the Irish Administration* (London, 1983), pp 37–8.

When the report was discussed in cabinet on 11 November,[81] the proposals were criticized on three main grounds: that Unionist Ulster had not opted for a separate parliament, but wished its citizens to be treated in all respects as British citizens; that the scheme gave the southern provinces less than the Act of 1914 offered, especially control of the police (which would be essential for effective administration), and hence would not be likely to prove acceptable to the Nationalists; that it would inflict hardship on the southern Unionists, and that since Sinn Féin was likely to dominate elections to the new parliament, their first action would be to declare a republic, unless this was provided against in some way by the bill.

However, it was generally recognized that these objections would apply to any home rule scheme for Ireland, and that repeal or postponement of the 1914 Act would be very undesirable, especially in view of relations with the dominions and the USA (the cabinet had before it letters in those terms from Viscount Grey, the former foreign secretary, now special ambassador to the US).[82]

The cabinet agreed that the best course would be to proceed by way of resolutions in parliament, embodying the main provisions of its scheme, which it asked the Long committee to work out in the form of a bill – especially the financial provisions.

At the next committee meeting (13 November) the Liberal, H.A.L. Fisher, played a leading role on two issues. He objected to the cabinet proposal, relayed by Long, to proceed by way of resolutions. Resolutions would be a waste of time, he said. Their only advantage would be that they might lead to alterations to the bill. What was needed to counter the 'general belief' in Ireland that the government did not keep its pledges was a just scheme, which would satisfy moderate opinion and would speedily be produced. (The committee agreed.) He also said that 'at the request of the prime minister' he had met Sir James Craig[83] and outlined to him the lines on which the committee was proceeding. Craig did not (as the cabinet had assumed) oppose the idea of a parliament in Belfast, but expressed himself against the inclusion of the whole province, and thought the six counties would be 'preferable'. 'His reason for this view was, of course, that the Protestant representation in the Ulster parliament would be strengthened and he also thought that the six counties would be a unit easier to govern than the whole province'.[84] The meeting ended with the chairman requesting that members would consider, before the next meeting, how the powers of the Irish Council might be enhanced.

This was the first occasion on which the Ulster Unionists officially accepted the necessity for a separate legislature in Belfast, instead of their previous inte-

81 CAB 23/20, 5 (19) (11 Nov. 1919). **82** CAB 23/20, 5 (19) (11 Nov. 1919). **83** Craig held junior office in the Coalition between January 1919 and April 1921 and so had ready access to ministers. **84** CAB 27/68, C.P.103 (13 Nov. 1919), p. 73.

grationist stance. Incidentally, two days earlier, Fisher reported to an informal conference of ministers that Craig was asking to be allowed to show the proposals 'in strict confidence' to Dawson Bates, secretary to the Ulster Unionist Council.[85] Long objected on the ground that if it were discovered that no final decision had been taken, it might cause 'agitation' to influence the committee, and the request was refused.[86]

When the committee submitted its fourth report to the cabinet (2 December),[87] it accepted that it had previously conferred such limited powers on the two Irish parliaments as would lead to administrative inefficiency, and so recommended that the powers previously transferred to the Council of Ireland, be exercized by both parliaments,[88] with the proviso that the parliaments could still by joint action transfer any of them to the council, if they so wished (customs were not included). In spite of Craig's strong advice, the committee still recommended separate parliaments for the province of Ulster and the three southern provinces, to be elected by PR. The committee also recommended that land purchase annuities be collected by the two Irish governments but transferred to the British exchequer, and that a free gift of the annuities for 1920 (£3 million) be handed over to get the new governments started.

The cabinet discussed the Irish question at great length on 3 December. Apart from the distraction of considering the report by the prime minister of a conversation with Lord Justice James O'Connor (allegedly representing the views of the Catholic hierarchy), who strangely opted for the separation of six counties on the ground that the separated counties would be attracted to reunification, since taxation would almost certainly be lower under an Irish government,[89] the cabinet's discussion centred on the Long committee report. They considered various options. The possibility of allowing the six counties to remain part of the United Kingdom under direct rule was speedily rejected. That proposal would meet with 'the strongest possible opposition' from the 'Covenanters',[90] would also be objected to by the Nationalists and anyway the administrative difficulties involved would be very great.

When the cabinet came to consider the possibility of two parliaments for the six and twenty-six counties respectively, a division immediately occurred between those, like Balfour,[91] who wished to keep Ulster, or at least the six counties, permanently separate from the rest of Ireland and the 'general feeling' that

---

85 CAB 23/20, 5 (19) (11 Nov. 1919).   86 CAB 23/20, 5 (19) (11 Nov. 1919).   87 CAB 24/94, C.P.247 (2 Dec. 1919).   88 Together with the following powers, which in the draft were reserved to the Imperial government for one year: local government, public health, housing, transport, agriculture, old age pensions, unemployment insurance and employment exchanges.   89 O'Connor asserted that he was more of a Catholic than an Irishman, and also admitted that the Irish people live in 'an atmosphere of delusions': CAB 23/18, 10 (19) (3 Dec. 1919).   90 Signatories to the Ulster Covenant who were not prepared to desert the Protestants of Cavan, Monaghan and Donegal.   91 Shannon, *Balfour and Ireland*, pp 250–1.

the ultimate aim of the government's policy in Ireland was a united Ireland with a separate parliament of its own, bound by the closest ties to Great Britain, but that this must be achieved with the largest possible support, and without offending the Protestants in Ulster.[92]

The majority considered the reasons for and against a separate parliament for the six counties, similar to those for and against the continued integration of the six. To the argument that the higher birth rate among Catholics would lead to the 'swamping' of the Protestants in the nine counties, statistics were produced to show the percentage of Catholics in Ulster had been declining for several decades (mainly through higher emigration rates), and then stood at 43 per cent. In view of the difficulty of getting the Covenanters to accept the scheme, and the superiority of the whole of Ulster as an administrative unit, the cabinet provisionally agreed that the bill should be drawn up for two parliaments, one for Ulster and the other for the three provinces. However, they added the proviso that if after the introduction of the bill it was found that the limiting of the northern area to the six counties would be more acceptable, the question could be looked at again. 'It was recognized, however, that the administrative problem would then require very careful examination.'

To the objection that the south and west of Ireland should not be given a block grant, because they had 'deserted Great Britain in its hour of need' during the war, the majority replied that since the Irish governments would not be empowered to impose anything except local taxation, a bonus had been proposed as compensation.[93]

> While some views were expressed in favour of keeping Ulster, or at any rate the six counties, permanently separate from the remainder of Ireland, the general feeling was that the ultimate aim of the government's policy in Ireland was a united Ireland with a separate parliament of its own, bound by the closest ties to Great Britain.[94]

At the next meeting to discuss the prime minister's speech to parliament the cabinet decided that he should present parliament with three possibilities for Ulster: (1) a nine-county parliament; (2) a six-county parliament; and (3) a six-county parliament with a boundary commission to draw the exact line of demarcation with a view to the inclusion of Protestant and Catholic communities, living near the border, within the jurisdiction of the appropriate parliament.[95]

---

92 CAB 23/18, C.10 (19) (3 Dec. 1919).   **93** Other matters, especially the financial arrangements, were shelved.   **94** CAB 23/18 (3 Dec. 1919).   **95** CAB 23/18 (16/19) (19 Dec. 1919).

Interestingly, in view of his subsequent attitude, the proposal for a boundary commission emanated from Sir James Craig, who was still lobbying ministers in favour of a six-county parliament. The government were 'inclined to lean towards' a six-county solution, and the proposal to set up a boundary commission, which at that time might produce 'considerable unrest' was deferred. The Long committee was divided on the question whether customs, excise and posts and telegraphs should be transferred to the new parliaments, and the matter was also shelved by the cabinet.

It was agreed that the prime minister would not commit the government on either excise or the postal service. Lloyd George himself believed that the retention of customs was a sign of unity.[96]

The prime minister introduced the resolutions leading to further legislation by the Commons 'in fine bravura style' (22 December 1919).[97] He began by pointing out that he was faced with the very difficult task of trying 'to compose a family quarrel'.[98] Three fundamental facts had to be faced: (1) the impossibility of severing Ireland from the United Kingdom; (2) the opposition of nationalist Ireland to British rule; and (3) the opposition of north-east Ulster to Irish rule.

In consequence, the government had to recognize that Ireland must remain part of the United Kingdom, while conferring self-government upon Ireland in all its domestic concerns, and establishing two parliaments, instead of one.[99]

The prime minister listed the services that would be reserved to the imperial parliament, some permanently (for example, the crown, defence, foreign affairs); others until the two parliaments agreed to unite (the post office, the higher judiciary, the police and the revenue departments). Land purchase would remain an imperial obligation, but each government would have to make an 'imperial contribution' of any surplus of revenue over expenditure.

Lloyd George emphasized the government's generosity by referring to the free gift to the parliament of the land annuities for that year, plus £1 million for establishment expenses, and pleaded for an end to recrimination and a 'fair consideration' of these proposals by Irishmen.[1]

The only other significant speech was made by Carson, who insisted that Ulster had thrived under the Union and should be allowed to maintain the *status quo*.[2] The Nationalists boycotted the discussion.

The reaction in Nationalist Ireland was, as Mansergh says, one of indifference born of a sense of irrelevance.[3] In January 1920 Walter Long visited Dublin and several towns in Ulster, and reported that he was surprised to find little interest

96 Jones, *Whitehall Diary*, iii, p. 39.   **97** Mansergh, *Unresolved Question*, p. 132.   **98** *Hans*, cxxiii (22 Dec. 1919), 1168. For the entire proceedings see 1168–85.   **99** *Hans*, cxxiii (22 Dec. 1919), 1168–85.   **1** *Hans*, cxxiii (22 Dec. 1919), 1185.   **2** Ibid., 1195. There was no formal debate. Lloyd George spoke on a motion for the adjournment.   **3** Mansergh, *Unresolved Question*, pp 133–4.

in the government's proposals in Ulster and bitter opposition to what they regarded as a betrayal from the southern Unionists.[4]

Just before the resolutions on the government of Ireland were moved, the IRA organized an abortive attempt to assassinate the viceroy, Lord French (19 December 1919). These two events set the pattern for 1920, 'the year in which British domination of Ireland ended'.[5]

4 CAB 24/97, C.P.571 (4 Feb. 1920).   **5** Mitchell, *Revolutionary Government*, p. 120.

6

1920

On 15 January 1920 occurred the first all-Ireland elections to be held under PR elections for boroughs and urban district councils. In the three southern provinces Sinn Féin won 560 seats; Labour, 394; Unionists, 355; home rulers, 238; independents and others, 269. Sinn Féin secured a majority in nine out of 11 corporations and 62 out of 99 district councils.

The results in Ulster were even more striking. In Belfast the Unionist block of 52 councillors was reduced to 29, with the Belfast Labour Party second with ten seats.[1] The total number of seats was 60 and the turn-out 60 per cent. Derry City for the first time returned a Nationalist majority.[2] Of the 573 seats up for election in the province 318 were won by non-Unionists, while the Unionists secured 255.[3]

In the other provinces the Sinn Féin total was not very impressive.[4] Immediately after the elections the Dáil local government department (under W.T. Cosgrave) asked all the councils with Sinn Féin majorities to pledge allegiance to the Dáil. Cork, Limerick and later Dublin did so, most of the others held back.

At this point it is appropriate to mention the county council election of June 1920, which resulted in an overwhelming victory for Sinn Féin. The party and its allies won control of 29 of the 33 counties.[5]

The second reading of the government of Ireland bill was due at the end of February 1920 but neither the cabinet committee nor the full cabinet had decided on the crucial questions of the size of the separate northern area (six or nine counties), nor whether the power of customs and/or excise would be transferred to the Irish parliament. So the prime minister[6] took the unusual step of asking Bonar Law, the Conservative leader and lord privy seal, to chair a special meeting of the Long committee.

Bonar Law reported that the committee, having carefully weighed the arguments on both sides, recommended that 'the whole of the province of Ulster should be included in the Northern Parliament'.[7]

---

1 Budge and O'Leary, *Belfast: Approach to Crisis* (London, 1973), pp 137–40.   2 The *Irish News* wrote (prematurely) of 'a death-blow to the Unionist clique' (19 January 1920).   3 Mitchell, *Revolutionary Government*, p. 124.   4 Ibid., p. 123. The first preferences were SF, 87,311; Unionists, 85,392; Labour, 57,626; home rulers, 47,102; Others, 44,273.   5 Tipperary had two county councils.   6 It has to be remembered that in 1919 Lloyd George had two other preoccupations apart from Ireland – the Versailles Conference at which he was the chief British delegate, and industrial unrest in Britain, especially among the miners.   7 CAB 27/68, C.D.664 (17 Feb. 1920).

On the question of customs and excise the committee was divided, so the decision was referred to the cabinet. Some (for example, Fisher) wanted to transfer both powers on the ground that otherwise the Irish parliaments' powers would be too limited; others (for example, Long) held that the power of customs was essential to sovereignty.[8] The committee referred the question of the judiciary to a sub-committee under Birkenhead, but were unanimous that land purchase was 'a debt of honour'[9] which the British government could not evade.

The full cabinet discussed the report a week later, with Lloyd George in the chair.[10] The cabinet considered the report in the light of the probable attitude of political parties so far as it had been ascertained', and concluded that the area of Northern Ireland should consist of the six counties plus two county boroughs (Belfast and Derry). The cabinet accepted the recommendation of the sub-committee that the two areas should each have a separate judiciary with appeals to the House of Lords. They agreed to defer the transfer of customs and excise until 'after the date of Irish Union'.[11]

This was the crucial meeting, which copperfastened the partition of Ireland. Why did the cabinet reject the decision of the cabinet committee in the previous week? Other authors have dealt with this question in various ways.[12] Although he is somewhat ambivalent, Mansergh obviously regards the decision as inevitable. It was 'unrealistic' to think of the whole of Ulster sustaining a separate parliament 'on so precarious a balance of political opinion'.[13] Boyce[14] does not advert to the discrepancy between the two meetings, but surmises that the Ulster Unionist demands had to be met because 'it would be difficult for the government to force through a scheme which was unacceptable to both their friends and to their critics'.[15] Lawlor makes the same point, but cites without comment the discrepancy between the cabinet committee and the full cabinet.[16]

It is difficult to argue that the full cabinet, with the same party balance as the committee, had any fresh information not available to the committee. What is more likely is that some members of both bodies changed their minds. The two most likely were the two former Conservative chief secretaries for Ireland, Arthur Balfour and Walter Long. Balfour, the philosophical politician, a supporter of the two nations theory, initially insisted that Ulster should continue to enjoy the 'noble privilege' of integration with the United Kingdom,[17] but by this date was lobbying the prime minister for a six-county parliament.[18] Long had previously argued

8 E.g. Irish Convention.   9 CAB 27/68, C.P.664 (17 Feb. 1920).   10 CAB 23/20, C.12 (20) (24 February 1920).   11 Ibid.   12 McColgan ignores the question.   13 Mansergh, *Unresolved Question*, p. 135.   14 D.G. Boyce, 'British Conservative Opinion, the Ulster Question and the Partition of Ireland, 1912–21', *Irish Historical Studies*, 17 (1970–1), p. 109.   15 This is a direct quotation from the cabinet Minutes of 19 December 1919, CAB 23/18 (16/19), p. 5.   16 S. Lawlor, *Britain and Ireland, 1914–23* (Dublin, 1983), pp 51, 236 n. 116.   17 Shannon, *Balfour*, pp 248–57.   18 On 10 February, Balfour wrote a personal letter to Lloyd George, regretting that 'the Cabinet opinion was turning against the small "Ulster to the large Ulster", which is, after all, no more than a geographical expression': Boyce, op. cit., p. 100.

for a nine county parliament on the ground – accepted by the majority of the committee – that it would be more likely to lead to reunification, and this would fit in with his federalist leanings. He is now recorded as reminding his cabinet colleagues of the force of the Ulster Unionist arguments for a six-county Ulster. He plainly would have preferred a nine-county area, 'which would provide a launching pad for a united Ireland', instead of a permanent partition, but he recognized that he had little choice if he wanted to save the government of Ireland bill.[19]

Shortly before the second reading of the bill the attitude of the Ulster Unionists was decided at a meeting of their governing body, the Ulster Unionist Council on 10 March. Carson urged the meeting to accept the six county system as the best available,[20] although he was conscious of the bitter opposition to what the Unionists of the three excluded counties regarded as 'desertion'.[21] The meeting was stormy. Lord Farnham, a landlord from Cavan, made an emotional speech urging rejection of the proposal. For a time it looked as if he might succeed, but by a show of hands the council accepted 'the throwing overboard' of Donegal, Cavan and Monaghan. One delegate was so outraged by the decision that she 'never cared for, or had anything to do with, politics from that day to this. Our Covenant was broken and loyal people were let down by us that day'.[22]

The second reading was introduced on 29 March by Ian Macpherson, his last public act as chief secretary for Ireland, since he would be replaced by Sir Hamar Greenwood on 2 April. Mansergh describes his speech as 'lack(ing) in conviction'.[23] It was virtually a mere recital of the provisions of the bill.[24]

The debate extended over three days.[25] The Coalitionists strongly supported the bill; five ministers spoke, ending with the prime minister. The speech by Bonar Law has been praised by his standard biographer as one of his finest parliamentary performances.[26] It was well constructed. Law reminded the House that the 1914 Home Rule Act was on the statute book and would automatically come into operation after the last Peace Treaty was signed.[27] So fresh legislation was urgently needed. He listed what he considered as the only four alternative policies, of which three were impracticable: repeal of the Act, because it would

---

19 Kendle, *Walter Long,* pp 188–9. See also D.G. Boyce, *Englishmen and Irish Troubles* (London, 1972), p. 109.   20 Jackson, *Carson,* pp 59–60.   21 The southern Unionists also accused the Ulster Unionists of desertion, but by now many influential southern Unionists were advocating dominion status for the whole of Ireland. See below, pp 81–3, 104, 107–8. The surviving diehards of the Irish Unionst Alliance who had rebelled against Midleton's compromise with Redmond, lacked both numerical strength and political weight.   22 *The Recollections of Mary Alice Young, née Macnaghton (1867–1946)* (Mid-Antrim Historical Group: 32, 1996), pp 61–2. The author was the grandmother of Rosemary, Lady Brookeborough.   23 Mansergh, *Unresolved Question,* p. 118.   24 *Hans,* cxxvii (29 March 1920), 925–44.   25 *Hans,* cxxvii (29 March 1920), 925–1036; (30 Mar.), 1107–218; (31 Mar.), 1289–336.   26 R. Blake, *The Unknown Prime Minister* (London, 1955), p. 418. Recently, Alvin Jackson has described the speech as 'a great success', in Robert Eccleshall and Graham Walker (eds), *Biographical Dictionary of British Prime Ministers* (London and New York), p. 270.   27 *Hans,* cxxvii (30 Mar.), 1122–8.

be in breach of the Coalition's pre-election pledges; a republic, which was out of the question; dominion home rule, which would inevitably lead to a republic, and would be incompatible with British security. The only feasible alternative was the bill. Law also referred to the 'fundamental fact', previously accepted by all 'leaders of opinion' in the House, 'that Ulster was not to be brought under a Dublin Parliament, except with her own free will'.[28]

Towards the end of his speech, Law asked rhetorically, 'In what conceivable way can having the six counties with a Parliament of their own make the union of Ireland more impossible than having the six counties as part of the United Kingdom? What is the difference?'[29] An answer had, in fact, been given by Captain Charles Craig (brother of James), who on the previous day had frankly admitted that he did not expect to see a united Ireland in his lifetime.[30] The reason was that the six counties would have a permanent Unionist majority. Referring to the 'heartbreaking' decision of the Ulster Unionist Council to ditch the three counties, Craig pointed out that those counties containing 260,000 'Sinn Féiners and Nationalists', with only 70,000 Unionists, would, if included, leave the Unionists with such a small majority that 'no sane man' would undertake to carry on a parliament in such conditions.[31]

All the Coalitionist speakers ignored the warnings from Craig, and professed their belief that the two Irish statelets would come together. Macpherson and Sir L. Worthington-Evans (minister for pensions) emphasized the generous financial provisions of the bill. The prosperity of Ireland had considerably increased since the War.[32] In 1912 public expenditure had exceeded revenue by £1.5 million, but in 1920 the country could afford to give back from the proceeds of the reserved taxes an 'imperial contribution' of £14.75 millions to the British exchequer.[33] The country would also receive as a 'free gift' £3.2 millions for land annuities in that year.[34] Although income tax, customs and excise were reserved, the Irish parliament would have control of new services, for example, old age pensions, housing, insurance and education.

The Irish Parliamentary Party had frequently claimed that Ireland had been overtaxed in the nineteenth century, and during the Treaty negotiations Michael

---

28 Ibid., 1123.   **29** Ibid., 1128.   **30** *Hans,* cxxvii (29 March 1920), 986–991.   **31** *Hans*, cxxvii (29 March 1920), 990–1. No member queried how Craig arrived at the figure of 260,000 Sinn Féiners and Nationalists in Cavan, Donegal and Monaghan. In the 1918 election the total vote for these parties was 39,087 in Donegal and 21,488 in Monaghan. The two Cavan constituencies were uncontested, but even if one were to assume that the entire 43,418 electorate turned out for Sinn Féin, the aggregate would fall far short of 260,000. Craig was simply indulging in a sectarian head-count of the entire population of the three counties.   **32** *Hans*, cxxvii (29 Mar. 1920), 925–36.   **33** After the first two years the imperial contribution would be levied by a new institution, the Joint Exchequer Board in the proportions of 56% and 44% between North and South. For the financial history of Northern Ireland after 1921 see R.J. Lawrence, *The Government of Northern Ireland: Public Finance and Public Services, 1921–1964* (Oxford, 1965).   **34** *Hans*, cxvii (29 Mar. 1920), 1033.

Collins was to demand a repayment of £400 million for overtaxation. Was this charge justified? A recent careful study has found the evidence inconclusive. On the one hand, a royal commission found in 1894 that 'Whilst the actual tax revenue of Ireland is about one eleventh of that of Great Britain, the relative taxable capacity of Ireland is much smaller and is not estimated by any of us as exceeding one twentieth'.[35] On the other, British policy at the turn of the century was to increase public expenditure in Ireland from £2.9 million annually in 1870 to £5.6 million in 1895. In 1907–8 even before old-age pensions swelled the total, it had risen to £7.8 million.

The main opposition speaker was Asquith (now returned to Westminster in a by-election). The former prime minister reminded the House that he had been a sponsor of the bill in 1893 and had introduced the bill of 1912. He pointed out that no one in Ireland wanted the 'costly and cumbrous' institution of two parliaments. He quoted with approval John Redmond's remarks in 1913: 'A unit Ireland is and a unit Ireland must remain'.[36] As to the provisions of the bill, apart from railways the Council of Ireland would have no powers at all, unless at a future date the two parliaments would delegate powers by identical Acts. Referring to Charles Craig's speech, Asquith asserted that by reducing Ulster to six counties, 'you have a majority which will enable you to defeat union'.[37]

Asquith's own policy would be to keep the 1914 Act, but amended to give Ireland 'to all intents and purposes' the status of a dominion. He would also provide within the Irish parliament an Ulster committee with special powers and functions. The area of Ulster would be determined by county option, 'the fairest suggestion'[38] – although he had not made this suggestion when he was prime minister.

The Asquithian Liberal and Labour speakers denounced the bill as denying the nationhood of Ireland.[39] The same point was made more forcefully by the Irish Nationalists. Their leader T.P. O'Connor pointed out that the bill would be passed without a single Irishman voting in its favour. Deriding the proliferation of public posts under the two parliaments, he stated: 'Since Jupiter wooed Danae with a shower of gold there has never been anything so bounteous as this bestowal of offices' by this bill.[40] Devlin reminded the House that Bonar Law had said that if Asquith 'dared' to put the Act of 1914 into operation he ought to be hanged on a lamp-post. (If so, so should Lloyd George, an author of that Act.)[41] The bill not only partitioned Ireland – but also Ulster; whereas the Nationalists had never agreed to a permanent partitioning. Might is right, he said, but only for a time.[42]

35 L. Kennedy and D.S. Johnson, 'The Union of Ireland and Great Britain, 1801–1921', in D.G. Boyce and A. O'Day (eds.), *The Making of Modern Irish History: Revisionism and the Revisionist Controversy* (London and New York, 1996), pp 43–7.   36 For the whole speech see *Hans*, cxxvii (30 Mar. 1920), 1107–19.   37 Ibid., 1113–14.   38 Ibid., 1118.   39 J.R. Clynes, the Labour leader, moved a hostile amendment, which was not pressed: *Hans*, cxxvii (29 March 1920), 944. 40 Ibid., 967–976.   41 *Hans*, cxxvii (30 Mar. 1920), 1134–52.   42 Ibid., 115.

In a somewhat embarrassed contribution Carson mirrored the dilemma of the Unionists. He could not and would not vote for Irish home rule, but would do nothing to prevent the bill from becoming law. ('Do you want me to go over and say to the Ulster people, Go and entrust your destinies ... to a Sinn Féin Parliament in Dublin?')[43]

Replying to the debate, the prime minister admitted that there was then no plan acceptable to any British party that would also be acceptable to any party in Ireland. The Irish were in danger of making the same mistake in relation to Ulster as the British had for 120 years made concerning Ireland.[44]

The second reading was carried by 348 votes to 94. Only the Coalitionists voted in favour, the opposition comprised the Asquithian Liberals, Labour, the seven Irish Nationalists, two southern Unionists (Dockrell and Jellett) and a handful of Tory 'die-hards'.[46] The Ulster Unionists abstained.

The committee stage of the government of Ireland bill was not taken until 10 May.

The success of the IRA campaign was demonstrated by the increasing rate of resignations from the police which by June was reaching over 250 a month. The campaign was helped by other factors: the appearance in November 1919 of a weekly propaganda sheet, the *Irish Bulletin*, which was regularly perused by British liberal editors, a strike by railway workers, refusing to handle imported war material (May 1920),[47] resulting in severe military inconvenience and most importantly, the success of Sinn Féin in the county elections of May 1920. Right from the start of the year the campaign was masterminded by the adjutant-general, Michael Collins,[48] who surpassed his colleagues in energy and decisiveness. Of the other IRA leaders, Cathal Brugha, the Dáil minister for defence, was originally friendly towards Collins, but grew to dislike him intensely even before the Treaty, when they took opposite sides, while Richard Mulcahy, chief of staff, greatly admired Collins.[49]

The British government were slow at first to recognize the IRA challenge; it was not adverted to during the second reading debate on the government of Ire-

43 Ibid. (31 Mar. 1920), 1289–99. **44** Ibid., 1323–36. **45** The Southern Unionist MPs, like the Ulster Nationalists, complained that the bill would leave them in a permanent minority in their area. *Hans*, cxxvii (30 Mar. 1920), 1158–170. **46** The Tory opposition included well-known dissidents, like Brigadier Page-Croft and Colonel Gretton, but also the senior member for Oxford University, Sir Charles Oman, who was a distinguished historian, but in politics an extreme reactionary. Later he voted against the Anglo-Irish Treaty and in 1928 opposed enfranchising women between 21 and 30, 'on principle'. **47** In March 1920 a group of armed men in disguise shot Tomas MacCurtain, lord mayor of Cork, and an IRA officer in his own house. The coroner's jury named two RIC officers as personally responsible as well as the lord lieutenant. In August two IRA volunteers at Collins' orders shot one of them, D.I. Swanzy, in broad daylight in Lisburn, where he had been transferred, precipitating a mini-pogrom of Catholic traders in a largely Protestant town. **48** See Coogan, *Collins*, pp 174–7. **49** For a recent affectionate tribute to Collins, see R. Mulcahy, *Richard Mulcahy, 1886–1971: A Family Memoir* (Dublin, 1999), pp 93–116.

land bill, but on 8 May Bonar Law told three ministers that 'Sinn Féin' had beaten the government and the situation was 'desperate'.[50] Meanwhile, the government had set great store by an inquiry by Sir Warren Fisher, head of the civil service, into the top administration in Ireland. Fisher's report was damning.[51] In an oft-quoted passage he wrote that, 'the Castle administration does not administer. On the mechanical side it can never have been good and is now quite obsolete; in the infinitely more important sphere (a) of informing and advising the Irish government in relation to policy and (b) of practical capacity in the application of policy it simply has no existence'.[52] Fisher strongly recommended that the 'almost woodenly stupid' top civil servants be replaced by 'suitable' officials from England. The resignation of the chief secretary, Macpherson, had taken place at the end of March, and he was succeeded by Sir Hamar Greenwood, a nominal Liberal home ruler, while the prime minister appointed General Sir Nevile Macready as commander-in-chief of the army.

The chief casualties of the Fisher Report were Sir John Taylor (assistant under-secretary) and Maurice Headlam (treasury remembrancer), two staunch conservatives. The under-secretary, James MacMahon, was retained mainly because he was a Catholic. The suitable English civil servants were headed by the chairman of the Board of Inland Revenue, Sir John Anderson,[53] later to become the supreme Whitehall mandarin who was made joint under-secretary with MacMahon, two assistant under-secretaries, A.W. (Andy) Cope and Mark Sturgis,[54] and four other officials loaned from London. The new team proved much more sensitive to Irish developments than their predecessors.

The turning point in British recognition of Sinn Féin growth was probably their success in May by winning control of 29 county councils, including Tyrone and Fermanagh, an achievement noted by English Liberal newspapers.[55] This success was followed by the Dáil instructing local authorities to refuse to pay the rates which they collected to the Castle government. Meanwhile the first suggestions of entering into discussions with Sinn Féin were made within the cabinet itself.

An important conference was held at 10 Downing Street on 30 April 1920. Those present included the prime minister, Bonar Law, Walter Long, and the top Irish officials Lord French, Hamar Greenwood and Sir Denis Henry (attorney-general). French gave an account of the position in Ireland. It was then a question, he said, either of war or a truce. 'They had declared war on us and it was

50 Fisher MS, 15, 3, 7 (8 May 1920).   **51** LGP, F/31/1/31 (12 May 1920).   **52** See J. McColgan, *British Policy and the Irish Administration, 1920–22*, pp 4–13.   **53** See below, p. 118.   **54** Sturgis at first was designated private secretary to Anderson out of deference to Cope's sensibilities. Sturgis (later Grant-Sturgis) left a racy set of diaries, recently edited by M. Hopkinson, *The Last Days of Dublin Castle: The Diaries of Mark Sturgis* (Dublin, 1999).   **55** Mitchell, *Revolutionary Government*, p. 126.

open to us to try to arrange a truce, to ask what the rebels wanted and to see if we were able to satisfy them'.[56] (The attorney-general pointed out that juries at assizes tended not to convict on murder charges and urged that more soldiers were needed, especially in four counties.)

This was the first occasion on which a truce was suggested at a ministerial conference. Since at a later meeting French opposed a truce, his suggestion may have been made to put the prime minister on the spot. But when pressed by French on whether to declare war on the rebels, Lloyd George forcefully replied: 'You do not declare war on rebels',[57] and stated that the disorder must be put down at whatever cost. A truce would be an admission of defeat.

On 2 June, Austen Chamberlain, soon to become Conservative leader, suggested that the cabinet should open negotiations with the Sinn Féin leaders in the hope of reaching agreement with them, but Lloyd George said that 'this would be making peace after a defeat'.[58]

The cabinet minutes for 2 June[59] indicate that a 'suggestion' was made 'that with a view to a comprehensive settlement of the present difficulties in Ireland, the time was approaching when the cabinet should consider the precedent of the 'Kilmainham Treaty' and the possibility of reaching a somewhat similar arrangement with the Sinn Féin leaders.[60] (However, the cabinet 'generally' agreed that before such a step was considered law and order would have to be restored in Ireland.)

The issue of a truce came up again at another conference on 23 July between the cabinet and the top Irish officials – French, Greenwood, Wylie,[61] Cope, Anderson, MacMahon, Macready and Tudor, being joined by Sir James Craig, still a junior British minister. Wylie[62] set the tone for the meeting by bluntly pointing to the virtually complete collapse of the court system in Ireland and predicting that within two months half the police force would have resigned through terrorism. He urged the introduction of dominion home rule for Southern Ireland and direct rule for the six counties. If the government were to make an offer on

---

56 CAB 23/21 (30 April 1920). This meeting was so secret that the minutes were marked 'not to be circulated widely'.   57 CAB 23/21 (30 Apr. 1920).   58 Churchill told his friend, Sir Henry Wilson: 'It was symptomatic that such a rotten proposal could be made': M. Gilbert, *Winston S. Churchill*, iv (1916–22), (London, 1975), p. 453.   59 CAB 20 (31) (2 June 1920).   60 CAB 31 (20) (2 June 1920). The 'Kilmainham Treaty' was an arrangement between the Gladstone government and Parnell (then in Kilmainham jail) that the latter would call off the agrarian agitation, which had landed him in prison, in return for the dropping of coercion by the government (May 1882). 61 W.E. Wylie (1888–1964), law adviser to the Irish government, later both a judge of the Irish Supreme Court (1920–1) and a judge of the High Court of the Irish Free State. T. Jones's editor K. Middlemass, confuses him with his uncle, J.O. Wylie, also an Irish judge, whose retirement in November 1920 left a vacancy which was filled by his nephew. Wylie retired from the Irish High Court in 1935 and like his colleague, James MacMahon, spent his later years as an officer of the Royal Dublin Society.   62 CAB 23/037, 336. Apart from the official record CAB 24/108, CP 1693 this conference is discussed at length in T. Jones, *Whitehall Diary*, iii, pp 25–31 and L. Ó Broin, *W.E. Wylie and the Irish Revolution, 1916–21* (Dublin, 1989), pp 84–98.

those lines he believed Sinn Féin would negotiate. Wylie was supported by all the Irish officials except French,[63] who kept silent, and Tudor who opposed. There was a clear conflict of opinion between the heads of the army and police; Macready siding with Wylie, while Tudor urged the militarisation of the RIC, the introduction of identity cards and other repressive measures.

Of the cabinet ministers, Churchill, Balfour, Bonar Law and Long all opposed the idea of dominion home rule. While emphasising that he was not in principle opposed to large concessions to Irish nationalism, Churchill refused to concede them at a time when they would be claimed as a victory for Sinn Féin. In the meantime[64] he favoured continuing with the Government of Ireland bill and raising 30,000 men in Ulster for military action throughout Ireland. Birkenhead pointed out that Wylie's proposal meant the abandonment of the bill. Balfour opposed dominion home rule on theoretical grounds; Ireland, he said, was not a nation and should be partitioned. (Was the dominion of Canada a nation?) Bonar Law claimed that there was no Irish Botha with whom the government could negotiate.[65] Long was horrified at the idea of dominion home rule; it would never be acceptable to the Conservative Party.

However, the essential irresolution of the cabinet was evident in that they were also planning coercive measures at the very time when a truce was being mooted. To counter the losses experienced through resignations by the RIC, the cabinet agreed to recruit large numbers of ex-servicemen at ten shillings a day (21 May).[66] These recruits, quickly nicknamed the Black and Tans because of their variegated uniform, were not, as rumour alleged, the sweepings of English jails, but rather 'the same type, and produced by much the same circumstances, as the Congo mercenaries of our own day'.[67] They quickly acquired a reputation for ruthlessness towards the civilian population.

On 27 July the cabinet decided on further reinforcements. Brigadier F.P. Crozier was a retired army officer, was appointed to raise the Auxiliary Division of the RIC, popularly to be known as Auxies.[68] These were ex-officers, hired at a pound a day. In time they acquired a similar reputation to the Black and Tans. The raising of the Auxies was welcomed by a new committee, the Irish Situation committee, set up on 24 June to assist the viceroy and chief secretary.[69]

---

63 On 27 July French wrote to the prime minister, stating that he had not spoken at the conference to give others a chance but he was in entire agreement with Tudor, Birkenhead and Churchill: L6P 48/6/37; also Ó Broin, *Wylie*, p. 99. Eunan O'Halpin has written: 'By the summer of 1920 French was little more than a mendacious blusterer, irrelevant to the administration of which he was technically still head': *The Decline of the Union: British Government in Ireland, 1892–1920* (Dublin, 1987), pp 206–7.   64 What follows is a conflation of the sources mentioned above, none of which is complete.   65 Law was resolved that the new system (the bill) should have its trial: R.J.Q. Adams, *Bonar Law* (London, 1999), p. 292. This is virtually the only reference to the bill in Law's latest biography.   66 CAB 20/30 (21 May 1920).   67 Lyons, *Ireland since the Famine*, p. 415. 68 F.P. Crozier, *Impressions and Recollections* (London, 1930) and *Ireland for Ever* (London, 1932). 69 CAB 23/21 (24 June 1920). The committee comprised Long (chairman), Balfour, Birkenhead,

The high-water mark of legislative coercion in Ireland was the Restoration of Order in Ireland Act (August 1920). The cabinet meeting on 26 July approved a request from the chief secretary to introduce a measure extending to Ireland the provisions of the Defence of the Realm Act 1916 – originally intended to cope with a war-time invasion.[70] The bill was preceded by a guillotine resolution – the first since 1918 – proposed by Bonar Law (leader of the House) limiting debate, on the grounds of 'necessity' to two days (2 and 5 August).[71]

Introducing the second reading, Greenwood admitted that a 'state of disorder' existed over a great part of Ireland with intimidation of juries, resignations of magistrates and failure of local authorities to perform their statutory duties. Accordingly, the government proposed to proclaim at their discretion certain areas in which the rule of law would be replaced by courts martial, empowered to impose the death penalty, while the government was also enabled to stop grants to local authorities defaulting on rates, and also to transfer prisoners to England. The debate resulted in the predictable conflict between British Conservatives supporting coercion and Irish Nationalists pointing to the failure of previous coercion bills. Perhaps the most interesting exchange was between Lloyd George and Asquith who questioned whether the administration of justice would be facilitated by this bill, while the government of Ireland bill was not supported by any section of Irish opinion. Asquith's remedy (which he admitted he had not advocated in the past) was the concession of dominion home rule which he believed would be acceptable to the majority of the Irish people. Lloyd George was at his most combative. Dominion home rule for Ireland would necessarily fail without control of an army, navy and the ports. He taunted Asquith with the charge that he would not propose unconditionally dominion home rule if he were still on the treasury bench. In Lloyd George's view there was no proposal the government could make which would be acceptable to the Irish people, except a republic which was out of the question. The last word in the second reading debate was that of Horatio Bottomley, who advised the House 'to face the facts and to recognise once and for all that British rule in Ireland has gone'.[72] The second reading was passed by 289 votes to 71.

On the committee stage on the second day (6 August)[73] Greenwood used the government majority to reject every amendment, even the sensible one of a time-limit to the bill in 'this farce of a debate'.[74] (It was pointed out that the Crimes Act of 1887 was still in force!) Eventually the great parliamentarian, Joe Devlin,

Churchill and Fisher. At its meeting on 22 July Churchill (secretary for war and air) admitted that the position of the troops was very bad, owing to their youth, and the strain caused by poor accommodation and the constant threat of attack. This admission was not recorded in the official minute (CAB 24/109, CP 1672). Townshend, *British Campaign*, p. 90.   **70** CAB 23/37(26 July 1920), 338.   **71** For the debates on the guillotine resolution and the bill, see *Hans*, cxxxii (5 and 6 Aug. 1920), 2689–722, 2723–963.   **72** Ibid., 2801.   **73** Ibid. (6 Aug. 1920), 2847–961.   **74** Ibid., 2883. Captain Wedgwood Benn (later Lord Stansgate).

lost his temper. He had complained about the absence of both the prime minister and the leader of the House (Bonar Law) and riled by persistent heckling from the Conservative benches, he delivered a stirring peroration: 'You may gather your majority, you may gather your profiteers, you may gather your corrupt politicians, and you may summon all your hosts, but you are not going to destroy the spirit of the nation, or the right of free men to speak their will.'[75] Devlin refused to sit down; he was named and ordered to leave the House, followed by the few remaining Nationalists. At the later stages of the bill the opposition numbers fell to 15. All stages were passed by 6 o'clock under the guillotine resolution. Several Conservatives voted for the bill with misgivings, including Lord Robert Cecil KC, who said the bill was treating Ireland as if it were a crown colony.[76]

Within a week of the passing of this coercion Act concern was expressed at a cabinet meeting that even if the measures were successful, there would be no party left in Ireland with which the government could negotiate – a tacit admission that there was then no 'constitutional' party left in the South.[77] The reaction in Ireland was not favourable. The previously staunchly Unionist organ, the *Irish Times*,[78] joined the *Freeman* and the *Independent* in urging an offer of dominion home rule instead of more coercion.[79]

The IRA too changed their tactics, going underground with 'flying columns', which could emerge out of nowhere and disappear into an at least not unfavourable environment.

There followed a succession of daring IRA attacks,[80] followed by 'authorized reprisals', which antagonized English liberals without reducing Republican activities.[81] Unquestionably the last months of 1920 witnessed the greatest increase in violence culminating in Bloody Sunday (21 November) when 12 intelligence officers were murdered in the morning followed by 17 civilians at a football final at Croke Park in the afternoon, and the raid in Kilmichael, County Cork, when 17 Auxiliaries were killed in an IRA ambush.[82] The following month saw the burning of part of the main street in Cork city, for which a subsequent inquiry established that an Auxiliary rampage was responsible.

The British government was aware of the breakdown of law and order but did not seem to appreciate the extent to which the clandestine Dáil was already establishing a counter-state.[83] After the county election results in June the local gov-

---

75 Ibid., 2912.   **76** Ibid., 2962.   **77** CAB 20/48 (13 Aug. 1920).   **78** *Irish Times* (4 Aug.1920). **79** W.E. Wylie was so outraged by the measure that he threatened to resign: *Sturgis Diaries*, i, pp 18–20.   **80** The cumulative totals of police and military killed from 1 January 1919 rose from 60 (30 June 1920) to 231 (31 Dec.): Townshend, *British Campaign*, (appendix V), p. 214.   **81** See C.P. Scott, *Diaries* (4 Nov. 1920), p. 389. Liberals and almost the entire British press agitated for the release of Terence MacSwiney, lord mayor of Cork, whose hunger strike lasted 74 days until his death on 25 October. MacSwiney was one of the first to be convicted under the Restoration of Order in Ireland Act.   **82** P. Hart, *The IRA and Its Enemies* (Oxford, 1998), pp 21–38; Townshend, op. cit., pp 129–31.   **83** For the development of the counter-state in 1920 see Mitchell, *Revolutionary Government*, pp 130–45.

ernment department ordered local councils to recognize the Dáil rather than the British government as the legitimate authority and cease to pay rate income to Dublin Castle. Within a month most county councils had obeyed. Sinn Féin courts were operating from May 1920, with considerable success outside the larger and towns and cities. In the West and South the courts were mainly agricultural, but in the East they had the services of some distinguished non-political persons, such as Alice Stopford Green and James Creed Meredith KC.[84]

However, the one part of the country where Sinn Féin was least effective was East Ulster, especially Belfast. To residents of the rest of Ireland Ulster was largely a terra incognita. Industrial Belfast resembled the South of Scotland more than the rest of Ireland. There was little travel between North and South. Ulster daily papers did not circulate outside the province and even in the ranks of the Irish Parliamentary Party only one Northerner, Joe Devlin, was prominent. Of the Sinn Féin members of the Dáil from Ulster only Eoin MacNeill and Ernest Blythe stood out.

Public attention turned to the North when, as so often previously, inflammatory speeches during the July celebrations led to the expulsion of 8,000 Catholic workers from the shipyards and engineering works and attacks on Catholic houses. The Belfast riots continued into September.[85]

In the meantime the Dáil was petitioned by Bishop MacRory of Down and Connor to organize a boycott of Belfast banks and goods. This the Dáil agreed with obvious reluctance in August, but confined the ban to banks and insurance companies. The boycott continued until the start of the civil war in 1922. It failed in the reinstatement of Catholic shipyard workers, but it undoubtedly reduced trade between North and South.[86]

The prospect of the passing of the government of Ireland bill did not greatly concern the Dáil republican leadership who naively expected it to be unworkable. The same opinion was held by Nationalist leaders in some parts of Ulster, for example, H.C. O'Doherty, mayor of Derry since January 1920, and T.J.S. Harbison, MP for Tyrone North-East, who believed that Nationalist opposition, active or passive, would leave the bill stillborn. But the most experienced Ulster Nationalist, Joe Devlin (MP since 1904), saw, as early as February 1920, that the bill was virtually certain to become law and would leave the Catholics in a permanent minority in the new six-county statelet.[87]

---

84 The Bar Council in Dublin passed a resolution prohibiting members from sitting in Sinn Féin courts. Meredith was one of the few barristers to do so.   85 By the end of the year 73 Catholics had been killed and nearly 400 wounded in Belfast: Eamon Phoenix, *Northern Nationalism*, p. 89. Other sources give slightly different totals.   86 Of the three Northern banks (the Ulster, Northern and Belfast) all lost at least 80% of their branches outside the province. Hardest hit was the Ulster Bank, 53% of whose branches were in the South, and which did not recover until the 1960s: D.S. Johnson, 'The Belfast Boycott, 1920–22', in J.M. Goldstrom and L.A. Clarkson (eds), *Irish Population, Economy and Society* (Oxford, 1981), pp 287–307.   87 Letter from Devlin to Bishop Patrick O'Donnell of Raphoe (13 Feb. 1920), cited in Phoenix, *Northern Nationalism*, p. 76.

Devlin was frustrated by the decision of the Irish Nationalist remnant to boy-cott the committee stage of the government of Ireland bill and felt inhibited from contacting ministers directly.[88] But no such inhibition affected Sir James Craig, rapidly becoming the real leader of the Ulster Unionists, who from his vantage point in Whitehall made frequent representations to the cabinet.[89] The framers of the bill intended to keep the seat of power in Dublin with the viceroy still pre-dominant and the government responsibilities largely undivided. Some functions were excepted (the crown, defence, foreign affairs, the higher judiciary), some were reserved until the two Irish parliaments would decide otherwise (customs and excise, income tax, the police and magistracy, the post office, the public record office, the registry of deeds).[90]

Before its introduction in February Ulster Unionists had threatened to oppose the entire bill if, as originally intended, there would be a single Irish judiciary.[91] On the committee stage Unionist opposition ensured that the police would be controlled by the lord lieutenant, and the two administrations. The post office, registry of deeds and public record office could be transferred to the Council of Ireland by agreement between the two parliaments, but land purchase would remain a British responsibility.[92]

The question of transferring income tax and customs and excise came up before a conference of ministers (only seven present) on 13 October. The Liberals, Shortt and Fisher, argued that the concession of 'full fiscal autonomy' would strengthen moderate opinion in Ireland. Surprisingly, they were supported by Walter Long, who argued that such a concession would secure wider acceptance of the bill.[93]

Ulster Unionists were opposed to the allocation of specific services to the Irish Council. They were even more opposed to the notion of a second chamber in the Ulster parliament. The bill as introduced provided for unicameral parliaments but the desirability of two second chambers to protect minorities was mentioned by Walter Long in the House of Commons on 18 May when he committed the government to the principle and promised a scheme for the report stage.[94] How-ever, by November it was clear that the Ulster Unionists would 'use the whole weight of their opposition' against such a scheme.[95] They held that the Northern minority would be sufficiently represented in a House of Commons (elected by PR). Accordingly, the cabinet referred to the Irish Council the duty to draw up a scheme for second chambers, subject to approval by both parliaments.[96]

---

88 Phoenix, *Northern Nationalism*, pp 89–90.   **89** To Carson, unlike Craig, the bill meant 'the painful desertion' of southern Unionists: Jackson, *Carson*, p. 60. Craig remained a junior minister until 1 April 1921.   **90** In the 1914 Act the post office, public record office and registry of deeds were transferred to Dublin. Their reservation in 1920 was for reasons of administrative conven-ience. McColgan, *British Policy*, pp 40–1.   **91** Accordingly Long had re-convened his commit-tee on 18 February to provide for two Irish judicial systems: Kendle, *Walter Long*, p. 188.   **92** McColgan, *British Policy*, pp 40–2.   **93** CAB 23/23 (13 Oct. 1920); Kendle, *Walter Long*, p. 195.   **94** *Hans*, cxxix (18 May 1920), 1267–8.   **95** CAB 23/23 (3 Nov. 1920), 61.   **96** Ibid.

The bill had already passed the committee stage with little change after the 13 October decision, but on 8 November Worthington-Evans (minister of pensions) moved to re-commit to include an amendment in terms of the cabinet decision on the Council of Ireland. Surprisingly, Carson supported the amendment on the ground that it would show that the Irish Council was not impotent.[97] The amendment passed by 175 votes to 31.[98]

On the report stage Worthington-Evans proposed two further amendments,[99] one providing that the two parliaments might by identical Acts vote to unite, the other, more ominously, providing that if in either parliament the number of members validly returned were less than half the total or if less than half had taken the oath within 14 days, the lord lieutenant might dissolve the parliament and rule with a committee of privy councillors, as in a crown colony.

The third reading was notable for another speech by Carson, lauding the bill as providing a procedure leading to 'a real unity and not a sham unity' in Ireland,[1] and a furious attack by Devlin on the introduction for the first time in a home rule bill of a second parliament, 'a permanent barrier against unity', and on the absence from the bill of adequate safeguards for the Northern minority. (He especially attacked the amendment introduced on the committee stage by the Ulster Unionists, allowing both parliaments to abolish PR after three years.[2] Devlin's colleague T.J.S. Harbinson of Tyrone showed the unreality of Nationalist opposition in the border counties by impotently threatening resistance to the partition scheme.[3]

When the bill went to the Lords the rift between Ulster Unionists and southern Unionists was plainly visible. The southern Unionist peers denounced the bill as a betrayal, leaving their 350,000 confreres to the mercy of a Sinn Féin dominated parliament of Southern Ireland.[4] Nor did they think much of the safeguards offered by the second chamber provision. One peer expressed a doubt that the council would ever meet. On the committee stage Lord Midleton, now the effective leader of the southern Unionist peers, proposed an amendment abolishing the 'fatuous' second parliament.[5]

When Midleton's amendment was defeated Lord Oranmore and Browne moved an amendment to insert a provision for a 'Senate of Southern Ireland' in the bill. Oranmore indicated that he favoured a senate on the model proposed

97 *Hans*, cxxxiv (8 Nov. 1920), 905–26. **98** On the committee stage of the bill the attendance rarely exceeded 200 out of 707 MPs. **99** *Hans*, cxxxiv (10 Nov. 1920), 943–7. **1** Ibid. (11 Nov. 1920), 1441. **2** Ibid., 1446–1462. **3** Ibid. (11 Nov. 1920), 1462. Devlin was also concerned that Craig in September had secured the agreement of the cabinet to re-establish the paramilitary Ulster Volunteer Force (UVF) as a Special Constabulary, since 'the Loyalists were losing faith in the Government's determination to protect them, and were threatening an immediate recourse to arms'. CAB 20 (48) (2 Sept. 1920), 49; Phoenix, *Northern Nationalism*, pp 93–4. **4** *Hans* (Lords), xlii (24 Nov. 1920), 596–638; 25 Nov., 630–8. **5** Ibid. (1 Dec. 1920), 799.

by the Irish convention, comprising government nominees, Catholic and Anglican bishops and representation of various socio-economic interests. Supported by Irish, Liberal and some Conservative peers the Oranmore amendment was carried by 120 votes to 36. Oranmore then moved an amendment for a Northern senate, carried without a division.

On the following day, following consultation with the Northern peers, Oranmore proposed that the Northern senate have ex officio the mayors of the two large cities and 24 to be elected by the 'lower House in such manner as that House may determine'. Another amendment by the former Irish under-secretary, Lord MacDonnell of Swinford, increasing the period during which PR would be compulsory from three to six years was narrowly passed on the report stage.

On 15 December the cabinet decided to accept the Lords' amendments concerning second chambers, but stipulated that the Northern senate be elected by PR. The cabinet rejected the MacDonnell amendment, largely because of Ulster Unionist opposition to PR.

Finally, a Lords' amendment entrusting Diseases of Animals to the Irish Council instead of the two parliaments was conceded with misgivings.

On the final Commons debate on the Lords' amendments Devlin and T.P. O'Connor repeated the criticisms of the bill's failure to protect the Northern minority. Neither was impressed with the amended provision for the senate. But to no avail. The bill quickly passed into law and received the royal assent on the last day of the session, 23 December 1920. It had failed to secure the votes of any Irish MP and was opposed by the entire Nationalist faction.

At the very time of the bill's last stages the Commons were voting (by resolution) to impose martial law on four counties. This had little or no effect on the IRA's activities. But at the very same time there occurred the first moves towards a truce in the war, which was becoming increasingly unpopular in Great Britain outside Conservative circles.[6]

A peace initiative occurred early in November, when two little-known persons, General Wanless O'Gowan and Dr W.M. Crofton, came to Ireland, interviewed republican leaders, including Griffith and Collins, and reported to Dublin Castle that the Dáil would accept the Government of Ireland bill, provided that they secured fiscal autonomy. Nothing further happened.[7]

Two more important peace initiatives at the end of 1920 were those of Father Michael O'Flanagan and Archbishop Peter J. Clune. O'Flanagan was vice-president of Sinn Féin and in the absence of de Valera the acting president.[8] At the begin-

6 C.P. Scott, *Diaries*, p. 389.   **7** The only known contemporary account of this initiative is in Sturgis, *Diaries*, pp 64–142, 249–51.   **8** O'Flanagan had been suspended from exercising his priestly functions by his bishop, because of involvement in politics (1918): Patrick Murray, *Oracles of God: the Roman Catholic Church and Irish Politics, 1922–37* (Dublin, 2000), p. 27.

ning of December he sent a public telegram to Lloyd George, reminding him that he had expressed willingness to make peace at once. He went on: 'Ireland also is willing. What first step do you propose?' Because of O'Flanagan's position[9] the cabinet decided to take the letter seriously and resolved (5 December) that the prime minister should write to O'Flanagan promising a safe conduct to any Dáil deputies who would wish to negotiate – in spite of the ban of 1919 – apart from 'certain individuals' who were 'gravely implicated in the commission of crime'.[10]

However, O'Flanagan's initiative was quickly repudiated. Padraig O'Keeffe, the secretary of Sinn Féin, pointed out that only the Dáil (of which O'Flanagan was not a member) had the authority to speak for Ireland and that the priest only spoke for himself. By the following month the cabinet accepted that O'Flanagan was not capable of negotiating with authority.[11]

A more significant initiative was that undertaken by Archbishop Peter J. Clune of Perth when visiting England in December.[12] Clune had connections with both sides; he had been chaplain-general to the Australian forces in France and he was uncle to Conor Clune, an IRA officer killed on Bloody Sunday. Clune was persuaded to intervene by some Irish in London, then met Lloyd George and was invited to go to Dublin and sound out the Sinn Féin leaders on the possibility of a truce. He duly met Collins, Griffith and MacNeill and reported to Lloyd George that they regarded a truce to be 'absolutely essential'. On hearing this, the cabinet decided to insist on the surrender of arms and ammunition as a condition of the truce (13 December). On 24 December the cabinet had a verbal message from Clune stating that Michael Collins ('the only one with whom effective business could be done') desired peace, but could not agree to the surrender of arms by the IRA.[13]

This was effectively the end of the Clune mission. Deeply disappointed, Clune told Andy Cope that if the negotiations had been in the hands of Anderson instead of the cabinet in London: 'We should now be in the fair way'.[15]

On his way back to Australia Clune called into the Vatican where he bluntly told Pope Benedict XV that a condemnation of violence in Ireland would be 'a disaster for the Church'. The condemnation was halted.[16]

---

9 'Even though his influence was minimal and he had attended only two of the twenty-three standing committee meetings which were held during the year': Michael Laffan, *The Resurrection of Ireland: The Sinn Féin Party, 1916–1923* (Cambridge, 1999), p. 290. **10** CAB 23/23 (6 Dec. 1920), C.60 (20); *Hans*, cxxxv (10 Dec. 1920), 2601–12. **11** Even though he persisted in meeting Lloyd George and Carson in December (Laffan, *Irish Revolution*, pp 290–1), and was even meeting Dublin Castle officials as late as February 1921 (Sturgis, *Diaries*, pp 127–9). **12** Clune was one of several Irish Catholic bishops in Australia at this time – Mannix (Melbourne) and Duhig (Brisbane), following Moran, appointed to Sydney in the 1880s. **13** CAB 23/23 (13 Dec. 1920), 70 (20). **14** CAB 23/23 (24 Dec. 1920), 77 (20). **15** Sturgis, *Diaries*, p. 105. Sturgis wrote (18 Feb. 1921): 'Speaking last night on the Clune truce talk the P.M. said that all his Irish advisers said truce without surrender of arms was impossible – that is contrary to my recollection and back pages bear me out ...': ibid., p. 129. **16** Coogan, *Michael Collins*, p. 202.

7

1921

> The will-o-the-wisp of sovereignty had been pursued at the expense of
> unity and in the end neither was served.
>
> John A. Murphy (on the Treaty delegates),
> *Ireland in the Twentieth Century* (Dublin, 1975), p. 38.

With the Government of Ireland Act on the statute book, the government immediately had to consider when the elections to the two new parliaments would be held. At a cabinet meeting on 29 December, attended by Macready and the other generals, the prime minister expressed the fear that Sinn Féin, which opposed the Act, might effect a boycott in the South by intimidation. He was reassured by the generals that 'the terror would be broken' within four to six months. On the other hand, the Unionists were anxious for an election in the North.[1] On the following day, the cabinet voted to extend martial law to four more counties, Clare, Kilkenny, Waterford and Wexford.[2]

But these optimistic hopes were not fulfilled. The extension of martial law, far from cowing the IRA, drove hundreds of young men into full-time guerrilla activity. At the beginning of the year the IRA structure was changed from brigades to divisions, with more authority to local commanders.[3] The repressive military measures: 'authorized reprisals' and internments (about 2,000) were accompanied by individual acts of brutality and indiscipline by the Black and Tans and Auxiliaries, which increased their unpopularity. Meanwhile, the clandestine Dáil remained active. Between its suppression in September 1919 and the end of 1920 only four sessions were held, but two sessions were held in January 1921, at which the Belfast boycott was extended to several classes of British goods.[4] At a third session (all private) on 11 March, the Dáil passed a decree prohibiting the census due in that year as 'a usurpation of the rights of the Irish people'. Dublin Castle decided not to proceed with the census, 'because it would be impossible to obtain complete and accurate returns'. So, for the first decade since 1801 there was no census for Ireland.[5]

---

1 CAB 79(A) (29 Dec. 1920).   2 CAB 81(20) (30 Dec. 1920).   3 For a recently published account of the activities of the East Clare Brigade under M. Brennan (later chief of staff of the army of the Irish Free State), see D. O Corráin (ed.), *James Hogan, Revolutionary Historian and Political Scientist* (Dublin, 2001), pp 39–40, 188–9.   4 Mitchell, *Revolutionary Government*, pp 244–7. 5 Ibid., p. 241.

On 26 January, Sir Edward Carson resigned as leader of the Ulster Unionists and on 4 February Sir James Craig was elected by the Ulster Unionist Council to succeed him.[6] In his farewell address to the council, Carson made this remarkable plea:

> Let us take care that we win all that is best amongst those who have been opposed to us in the past in this community. Let us show that while we were always determined to maintain intact our own religion and all that it means to us, we consider that they have a right to expect that all that is sacred to them in their religion will receive the same toleration.[7]

These sentiments were not to be shared by the new Unionist leader.

On 18 February Brigadier General Crozier resigned as commander of the Auxiliaries after just six months in office. He later wrote to *The Times*, citing his disgust at the indiscipline and rapacity of the force.[8] What actually precipitated his resignation was the discovery of a plot by the Auxies to murder Bishop Fogarty of Killaloe (treasurer of Sinn Féin) and dump his body in the Shannon. Crozier sent an anonymous warning to the bishop, who immediately left Ennis.[9]

One may wonder what sane person would expect any conceivable gain to follow the murder of an Irish Catholic bishop – the first since 1681?[10] It does not need to be stressed that in the Ireland of the 1920s Catholic priests were universally respected, while the bishops, the successors of the Apostles, were regarded as the 'Oracles of God'.[11] The outpouring of public grief and execration of the British forces that would have followed the murder of Bishop Fogarty can well be imagined.

Meanwhile, public opinion in England was becoming hostile to the government's Irish policy. At the end of January, a Labour Party commission of inquiry concluded that 'Things are being done in the name of Britain which must make our name stink in the nostrils of the whole world'. The archbishop of Canterbury condemned the government in the House of Lords, and Lord Robert Cecil and another Conservative MP took a similar line in the Commons.[12]

During these months two senior ministers, both hardliners on Irish policy, resigned on grounds of ill-health.[13] Walter Long resigned on 12 February and was given a peerage. On 17 March Bonar Law resigned, both as lord privy seal and Conservative leader, being succeeded in the latter post by Austen Chamberlain, who had been the first cabinet minister to suggest an approach to Sinn Féin.[14]

---

6 *Belfast News Letter* (5 Feb. 1921).   **7** Carson was appointed a lord of appeal on 21 May 1921; Craig continued as financial secretary to the admiralty until 1 April 1921.   **8** *The Times* (8 April 1921).   **9** This story was first told in F.P. Crozier, *Impressions and Recollections* (London, 1930). On reading it, Fogarty wrote thanking Crozier for his 'humanity'. The letter was reproduced in an appendix to a later book, *Ireland for Ever* (London, 1932), p. 300–1.   **10** That is, St Oliver Plunket.   **11** P. Murray, *Oracles of God*.   **12** Jones, *Whitehall Diary*, iii, pp 48, 53.   **13** Law, suffering from exhaustion, was peremptorily ordered by his doctors to take six months complete rest. It is unclear whether he intended at this time to resume political activity, as happened eight months later. Adams, *Bonar Law*, pp 294–8.   **14** See above, p. 101.

Perhaps misled by military optimism, the prime minister set the date for both Irish parliamentary elections as 24 May. However, even in Dublin and the martial law areas the IRA remained active. On 5 March Lord Midleton, speaking with the authority of a former secretary for war, assured ministers that 'Whatever any soldier has said to the contrary, no civilian that I have met will admit that we have gained in the last six months',[15] and urged the postponement of the elections on the ground that it would be impossible to ensure a free election in the South. (March 1921 was the worst month for outrages since November 1920.)

Midleton's was the first of many efforts to postpone the elections, at least for the parliament of Southern Ireland; but the government had initiated a process that was to prove unstoppable. Even before the date was fixed, politicians were preparing for the first ever parliamentary elections all over Ireland under the single transferable vote system of PR. Since more than any other event the May elections affected the future course of Irish politics, they deserve detailed treatment.[16]

ELECTIONS TO THE PARLIAMENT OF SOUTHERN IRELAND

In January 1921 Sir Hamar Greenwood[17] expected that Sinn Féin would lose seats in the Southern elections, but a colleague in the Irish administration noted that Sinn Féin was confident that it would 'sweep the lot'.[18] At first, there were some indications that non-Sinn Féin candidates would emerge. Dr Ashe, a Southern Unionist, promised Dublin Castle a number of 'moderate' candidates and thought they might capture 15 to 20 seats, if he had the money. But none emerged.[19] More remarkably, the two non-Sinn Féin members elected in 1918 for territorial constituencies – Captain William Archer Redmond (Nationalist, Waterford) and Sir Maurice Dockrell (Unionist, Dublin, Rathmines)) stood aside. A former Dublin Nationalist councillor, Dr J.C. MacWalter, intended to stand on behalf of a stillborn Centre party, but was dissuaded. It soon became plain that the only potential challenger to Sinn Féin would be the Irish Labour Party, which had abstained in 1918, but after discussions with Sinn Féin the only concession was that Richard Corish, a Labour councillor in Wexford town, was included in the Sinn Féin panel for the Wexford four-seat constituency.[20] So for the second time the Labour Party abstained from a general election outside Belfast.

15 Jones, *Whitehall Diary*, iii, p. 54.   **16** The definitive study of the elections to the Northern parliament is Sydney Elliott, 'The Electoral System in Northern Ireland', i, pp 153–76, a PhD thesis for Queen's University, 1971, which is unfortunately still unpublished. There is a brief account of the election to the Southern parliament in Laffan, *Resurrection of Ireland*, pp 336–41.   **17** Tim Healy's acerbic comment on Greenwood (27 April 1921) was 'Out of Bedlam no such councillor of the King as Greenwood could have been selected': Frank Callanan, *T.M. Healy* (Cork, 1996), p. 558.   **18** *Sturgis Diary* (13 Jan. 1921), p. 110.   **19** Sturgis regarded this pledge as 'Moonshine': ibid. (9 Mar.), p. 139.   **20** In 1922 Corish returned to his old allegiance and as a Labour candidate easily headed the poll in the Wexford constituency. He remained a TD until his death in 1945, when he was succeeded by his son, Brendan, later leader of the Irish Labour Party.

By 1 April it was clear that Sinn Féin would contest all the constituencies, except Trinity College, Dublin, and no opposition had emerged. De Valera drafted the party manifesto and assured voters that a vote for Sinn Féin would legitimize the republic.[21] Since the PR system merged existing single-seat areas into multi-member constituencies, ranging from Cavan (three-seat) to Kerry-Limerick (eight-seat), the president instructed existing constituency executives to unite for the purpose of the election. As might be expected, almost all existing members of the First Dáil were re-selected. Seventy-five seats had been won in 1918, but there were only 71 members, since de Valera, Griffith, Professor Eoin MacNeill and Liam Mellows had each been returned for two constituencies. Since then Terence MacSwiney (Cork Mid) and Pierce McCann (Tipperary East) had died and three had resigned:[22] the Sligo executive did not re-nominate J.J. Clancy in spite of protests from headquarters. The remaining incumbents were selected.

The 57 new candidates have been referred to as 'politicians by accident', but they contained several who were to hold ministerial office: Patrick Hogan and General Sean MacEoin were to become ministers in Cumann na nGaedheal and Fine Gael coalition governments; Patrick Ruttledge, Thomas Derrig and Sean Moylan were to be Fianna Fail cabinet ministers. General Eoin O'Duffy would be successively chief of staff of the Free State army and commissioner of the Gárda Siochána. Erskine Childers would be appointed secretary to the Treaty delegation. Lastly, at a time when there was only one female member of the House of Commons (Lady Astor) and none in the French chamber of deputies, Sinn Féin nominated six women to serve in the Southern parliament.[23]

Sinn Féin appointed Austin Stack as director of elections. He was the least efficient member of the cabinet, but his task was easy.[24]

By the third week of April no alternatives to Sinn Féin had come forth and one candidate wrote in his diary that 'there is not the shadow of a shade of a sign' of an election.[25]

On 27 April, a chastened Greenwood told the cabinet that he had been much too optimistic about 'breaking the terror' and could not now give a date when that would occur.[26] It was evident that all the Southern constituencies would be uncontested and he asked whether the elections could be postponed, as the southern Unionists wished? Nobody argued for postponement. It would be an admission that the Act had failed

21 Laffan, *Resurrection of Ireland*, p. 335.   **22** Roger Sweetman (Wexford North) had resigned through opposition to IRA policy.   **23** Mrs Margaret Pearse, Mrs Kathleen Clarke, Mrs Katherine O'Callaghan and Mary MacSwiney were related to heroes of 1916 and the war of Independence. Countess Markievicz was re-nominated. Dr Ada English represented the National University of Ireland.   **24** Michael Collins' remark: 'Austin, your department (Home Affairs) is a bloody joke' is well known.   **25** Liam de Roiste (Sinn Féin, Cork Borough), 25 April 1921, cited in Laffan, *Irish Revolution*, p. 339.   **26** Jones, *Whitehall Diary*, iii, gives a detailed account of this meeting, and of the meeting of 12 May (pp 55–70). See also Fisher Diaries, 21 and 27 April, 11 and 12 May 1921; CAB 21 (39) (12 May 1921).

except for the Northern elections. Also, it could be alleged that that was the real goal of the cabinet. But if the elections were held and Sinn Féin monopolized the new parliament and refused to take the oath to the king, then the parliament could be dissolved and Southern Ireland be governed like a crown colony. The Liberal minister, Edwin Montagu,[27] was appalled at that prospect, and he led the Liberal ministers in making a strong case for a truce and negotiations with Sinn Féin, since they were clearly the dominant Irish party. The existing situation was 'degrading to the moral life of the whole country' and creating an odious impression abroad. A new effort would put the onus on Sinn Féin and make it difficult for them to resume the war.

The Conservative view was articulated by Balfour. A truce would be a sign of weakness. One did not negotiate with murderers. Others were less sure and some (Chamberlain and Curzon) changed their minds. Lloyd George was decisive as usual. If a truce were followed by negotiations, the Irish would surely ask for more – perhaps not a republic, but even dominion status with full fiscal autonomy would be unacceptable. Irish workmen would be paying less for tea and tobacco than the English. His attitude was that the British could not go further than the 1920 Act. 'These people will come round sooner or later.'[28] Greenwood was not present at the meeting, but had with the Irish generals sent letters opposing a truce. Anderson and his immediate subordinates were in favour of a truce and an offer of dominion home rule for most of Ireland.[29]

Lord FitzAlan (formerly Lord Edmund Talbot) had been appointed viceroy in April and a special Act had been rushed through parliament to enable him, a Catholic, to take up the post. His attitude was 'Let Sinn Féin take the initiative'. When the cabinet voted, all the Liberals, except Lloyd George and Shortt, were for a truce, all the Conservatives against. The proposal was defeated by nine votes to five. The formal conclusion was 'that it would be a mistake for the government to take the initiative in any suspension of military activities in Ireland, and that the present policy in that country should be pursued'.[30] Fisher wrote ruefully: 'We lose the day.'[31]

On 6 May John Dillon issued a statement, asserting that the British repressive policy made it practically impossible for any Nationalist Irishman to contest the election for the Southern parliament.[32] This was the swan-song of the party that had contested and dominated the Irish parliamentary representation between 1874 and 1918. By nomination day (13 May) the Southern election was over. Sinn Féin candidates were unopposed in all the 26 territorial constituencies and the National University

---

27 CAB 39(21) (12 May 1921).   **28** Jones, *Whitehall Diary*, iii, pp 63–70; CAB 39(21) (12 May 1921).   **29** On 3 March 1921 the *Irish Times* published an anonymous letter from Mark Sturgis, urging the British government to offer dominion home rule. This was followed by a similar anonymous letter (21 April) from Richard Southwell Windham-Quin, master of the horse to the lord lieutenant, a friend of Sturgis, known in later decades as Lord Adare, a leading racehorse owner in Co. Limerick, who succeeded his father as 6th earl of Dunraven in 1949, died 1965.   **30** CAB 39(21) (12 May 1921).   **31** Fisher Diaries (12 May 1921), p. 68.   **32** *Freeman's Journal* (9 May 1921). This election is not mentioned either in Lyons, *John Dillon*, or Callanan, *T.M. Healy*.

of Ireland (four seats); four independent candidates were returned unopposed for the Dublin University (Trinity College) constituency.[33] Thus the parliament of Southern Ireland became the first (and hitherto the only) democratic parliament to be entirely elected without a contest.

ELECTIONS TO THE PARLIAMENT OF NORTHERN IRELAND[34]

The Sinn Féin cabinet did not express any interest in the fate of the Government of Ireland bill in 1920, affecting to believe that it would be unworkable in both parts of the country. However, at the beginning of 1921 they were shaken out of their complacency by a warning from a Northern Nationalist source that the Northern parliament would soon be a reality. Divided counsels emerged, de Valera arguing that unless Sinn Féin was assured of winning at least 10 seats it should boycott the Northern election. Collins more realistically urged that Sinn Féin contest the Northern seats in line with its previous electoral policy, but on a strict understanding that they would not attend the parliament. (He expected 13 or 14 seats.) The Dáil cabinet agreed, but it was obvious because of the PR system that the party would have to do a deal with the Nationalists to maximize their performance. Devlin was agreeable and after much discussion a pact was arranged by Nationalists and Sinn Féin: each party would nominate up to 21 candidates, and supporters of each would be urged to give their second preferences to the other.[35] Devlin disliked the 'fatal' policy of abstention, but de Valera insisted that both parties commit themselves to it.

The pact, signed by de Valera and Devlin, was approved by an 800–strong Nationalist conference in Belfast on 4 March. The conference resolved to record its 'unalterable belief' in the right of Ireland to self-determination, to regard the prospect of a parliament for a section of the province of Ulster as 'a menace to public unity', and that all candidates pledge themselves neither to recognize or enter into it.[36]

To Devlin's dismay, outside Belfast the Nationalist organisation was largely moribund and the party was able only to field 13 candidates. Of the four Nationalists elected for six-county seats in 1918 only Devlin and T.J.S. Harbison stood again[37], the remaining candidates, apart from T.J. Campbell KC and John D. Nugent (national secretary of the AOH and former MP), were undistinguished.[38]

Sinn Féin produced 19 candidates, a measure of their self-confidence. De Valera (Down), Collins (Armagh), Griffith (Fermanagh-Tyrone) and MacNeill (Derry) stood

33 Professors W.E. Thrift and E.H. Alton, both subsequent provosts, Professor Sir James Craig (a surgeon) and Gerald Fitzgibbon KC (later a Supreme Court justice in the Irish Free State). **34** For this election see Elliott, 'The electoral system in Northern Ireland', i, pp 153–76 and Phoenix, *Northern Nationalism*, pp 106–32. **35** At first the unanimous opinion of Devlin's party officials in Belfast was in favour of boycotting the election, Phoenix, *Northern Nationalism*, p. 116. **36** Elliott, 'Electoral System', pp 155–6. **37** Patrick J. Donnelly (Armagh South) and Jeremiah McVeigh (Down South) did not stand in 1921. **38** Campbell was a former editor of the *Irish News* and on Devlin's death in 1934 would become leader of the Nationalists in the Northern parliament.

again. So did two other TDs, Sean O'Mahony (Fermanagh-Tyrone) and Sean MacEntee (Belfast West).[39]

On the eve of the election Bishop MacRory of Down and Connor issued a pastoral, urging all Catholics, whether republican or constitutionalist, to vote together, since the future of Catholic education was at stake.[40]

The Unionists were also aware of the fissiparous tendencies of PR and urged the electorate to poll 'the full Unionist strength' in what they regarded as a plebiscite on their new constitution, the government of Ireland Act. Labour and independent candidatures were condemned as playing into the hands of the opposition. On nomination day only five independents out of 77 candidates stood (40 Unionist, 19 Sinn Féin and 13 Nationalist).

During the campaign occurred the celebrated trip of Craig to Dublin to meet de Valera. There was no specific agenda and nothing was decided. Afterwards the Unionist press claimed that the meeting was arranged at de Valera's request. In fact, Andy Cope was the instigator.[41]

As polling day approached, the electoral polarisation increased, with fervent appeals on both sides for or against the 1920 Act. The results surprised even the Unionist hierarchy. On a turnout of 89 per cent, all the Unionist candidates (40) were elected – a better result than they were ever to achieve under the majority system, to which they reverted in 1929. Nationalists and Sinn Féin secured six seats each, well below their expectations.

Immediately the Nationalist press wailed about 'gerrymandering' – a complaint not previously made during the campaign.[42]

Subsequent writers have claimed that the electoral boundaries were rigged to suit the Unionists.[43] This allegation seems unfair. Outside Belfast the constituencies were the counties, as they had been throughout the nineteenth century.[44] Belfast was divided into four divisions, as in 1885–1918. The real answer to the charge of gerrymandering in 1920 is the result of the election of 1925.[45] Under the identical electoral system and boundaries the turnout fell to 75 per cent, and the Unionists lost eight seats, including those of two ministers. The gainers were Labour, Independent Unionists, Independents and Nationalists.[46]

The election results of 1921 confirmed what was already known, that the Unionists would dominate the Northern parliament, and Sinn Féin the Southern. But while there were some non-Unionists in the Northern body – although they were not to take their seats – there would be only four non-Sinn Féin members in the Dublin par-

39 O'Mahony was the only Dáil deputy to be elected solely for a constituency in Northern Ireland.   40 Phoenix, *Northern Nationalism*, pp 128–9.   41 Elliott, 'Electoral System', pp 161–6; *Sturgis Diary* (5 May 1921), pp 170–1.   42 *Irish News* (25–28 May 1921); Elliott, 'Electoral System', pp 166–74; Phoenix, *Northern Nationalism*, p. 131.   43 Coogan, *Collins*, p. 212; Laffan, *Resurrection of Ireland*, p. 340.   44 There was virtually no cross-voting on the Unionist side, Elliott, 'Electoral System', pp 173–6.   45 Ibid., pp 298–304.   46 The Nationalists, for the first and only time, secured a seat in Co. Antrim (a seven-seat constituency) in 1925.

liament (from the Dublin University constituency). So Sinn Féin won every territorial seat in the parliament. This result is unique in the history of democratic elections, but remarkably has not been emphasized by historians, who have reported the results without comment. But to contemporaries the scale, as well as the fact of the Sinn Féin victory, was evident.

On 14 May 1921 the *Irish Times*, which had consistently supported the Unionist cause, produced a remarkably prescient editorial:

> The Southern elections have put *Sinn Féin* in a position of indisputable strength as the spokesman of a large majority of the people. They have created a popular assembly which, though it refuses to become a Parliament, must be the country's brain and voice for all purposes of political negotiation.[47]

Such clear-sightedness was not found in the British government.

At its meeting on 24 May[48] the cabinet set dates for the opening of the two new parliaments – 21 June for the Northern and 28 June for the Southern parliament. Realizing that the Sinn Féin deputies would boycott the Southern parliament and constitute themselves as the Second Dáil,[49] meeting in secret, the cabinet ruled that in that case the twenty-six counties would have to be ruled as a crown colony under martial law. The Irish Situation committee, which had lapsed since August 1920, was re-established with the specific instruction to make detailed preparations for martial law. Military reinforcements should also be provided for Ireland.

At its meeting on 25 May, anticipating a further Craig–de Valera encounter, the cabinet decided that Craig could be informed by Lloyd George that the cabinet would not assent to any proposal to permit the Irish parliament to impose customs duties. In an unusual move three ministers (Montagu, Churchill and Fisher) recorded their dissent.[50]

The Irish Situation committee held its first meeting on 26 May.[51] It approved the introduction of martial law on 12 July if the Southern parliament did not function. The minutes made the revealing admission that the extension of martial law 'would obviate the necessity of authorized reprisals, which are at present resorted to as a safety valve for the soldiers and police'. But there was also an admission that the English people generally were against resorting to martial law, except in extreme cases.

As to the morale of the military, Macready, the commander-in-chief, sent a gloomy assessment to the committee. Many of the soldiers had to put up with two nights sleep per week. If the present situation continued beyond the summer, 'steps must be taken

47 *Irish Times*, 11 May 1921.   **48** CAB 21(41) (24 May 1921).   **49** The first Dáil held four meetings between January and May 1921, all in private, using various houses in Dublin: Mitchell, *Revolutionary Government*, pp 347–8.   **50** CAB 21(42) (25 May 1921).   **51** CAB 27(107) (26 May 1921). The committee comprised Austen Chamberlain (chairman), Shortt, Churchill, Worthington-Evans, Balfour, Fisher and Greenwood.

to relieve practically the whole of the troops, together with the great majority of their commanders and their staffs'.[52] In an accompanying memorandum, the secretary for war admitted that there was a risk of virtual stalemate throughout the summer, and autumn and winter 'will be a time of decisive advantage to the rebels'.[53]

Meanwhile, IRA activities continued unabated, even in the martial law counties, their most spectacular (and deplorable) achievement being the burning of the Dublin Custom House on 25 May, destroying centuries of public records.[54]

There is a surreal quality about the proceedings of the Irish Situation committee on 15 June.[55] They discussed line by line Macready's forthcoming proclamation. 'No state of war exists,' they proposed; 'what does exist is an insurrectionary movement supported by murder'. They took no account of the fact that the Dáil comprised the elected representatives of the people, but gravely considered whether to proclaim the body and arrest all the members. Greenwood was all for trying de Valera and Griffith for treason. Macready said that de Valera must be tried for his life; otherwise the troops and police 'would regard the proceedings as a farce'. Anderson pointed out that mere membership of the Dáil could not be regarded as treason, but only active membership. He personally was against treating the Dáil in the same way as the IRA or IRB. Macready actually imagined that the proclamation of the Dáil would encourage the 'moderates'. Shortt urged that TDs should be given a date to withdraw from that body before being proceeded against. What was unsaid was that such trials would have to be by military courts, since civil judges would not touch them.

Alarmed by the hawkish tone of the government, Anderson,[56] departing from his role as a meticulous civil servant, wrote a personal letter to Greenwood, with the request that it be shown to Chamberlain. If the government decided on martial law, they should first 'announce the extreme limits of concession to which they are prepared to go in the direction of Dominion Home Rule'. It would be 'the wildest folly' to embark on such repression without the conviction that they would be supported by parliament and the country.[57]

Meanwhile, preparations were being made for the formal opening of the Northern parliament on 18 June. Following the elections, Sir James Craig had been sworn in as prime minister with a small all-Unionist cabinet, and in advance of the opening of parliament Craig persuaded Whitehall to lend them a senior civil servant to establish the new Northern civil service[58] and invited King George V to perform the opening ceremony, an invitation which the cabinet (with some misgivings on the

52 CAB 27(107) (25 May 1921) 429.   53 Ibid., 427.   54 Between May and 11 July there were 162 police and military casualties – over one quarter of all crown casualties since the beginning of 1919: Boyce, *Englishmen and Irish Troubles*, p. 185. See also D. Fitzpatrick, *Politics and Irish Life, 1913–1921: Provincial Experience of War and Revolution*, pp 225–31.   55 It was a small meeting: Chamberlain, Shortt, Balfour, Greenwood, with Macready.   56 ACP 31/2/3 (18 June 1921).   57 ACP 31/2/3 (18 June 1921). Anderson's solution was dominion home rule for all Ireland, provided the violence ceased: J.W. Wheeler-Bennett, *John Anderson, Viscount Waverley* (London, 1962), p. 70.   58 For the setting up of the Northern civil service see McColgan, *British*

ground of security) approved. At the beginning of June there arrived in Britain the only mediator in the Anglo-Irish dispute who was even partially successful, General Jan Christiaan Smuts, prime minister of the Union of South Africa and previously rebel against British rule.

Smuts lunched with the king on 13 June and found him disposed to make a conciliatory speech during his visit to Belfast. On the following day Smuts wrote to Lloyd George.[59] The present situation in Ireland, he wrote, was 'an unmeasured calamity'. Not only did it negate all the principles of government, 'which we have professed as the basis of the empire', but the present methods are 'frightfully expensive' and they have failed. The establishment of the Ulster parliament would definitely eliminate the coercion of Ulster, and the road was clear for a more statesmanlike approach to the rest of Ireland.

Smuts suggested that the king take advantage of the ceremony in Belfast to foreshadow the grant of dominion status to the rest of Ireland. Such an offer, coming from the king, might be acceptable to the Irish leaders and informal negotiations could then be set going.[60] Smuts enclosed a proposed draft for the king's speech.

The cabinet objected to the 'gush' in Smuts' draft. Other drafts were proposed, but the final draft to be used by the king on 18 June was written by Sir Edward Grigg, private secretary to the prime minister.

The speech was sensational. The king made only a perfunctory reference to the new institutions[61] in Belfast and then continued in this famous passage:

> I speak from a full heart when I pray that My coming to Ireland to-day may prove to be the first step towards an end of strife amongst her people, whatever their race or creed.
>
>   In that hope I appeal to all Irishmen to pause, to stretch out the hand of forbearance and conciliation, to forgive and to forget, and to join in making for the land which they love, a new era of peace, contentment and goodwill.[62]

The reaction to the king's speech was overwhelmingly favourable, and on his return from Belfast he was cheered on the streets of London. It was in a changed mood that the cabinet met on 24 June.[63] Lloyd George reminded his colleagues that the king's appeal for reconciliation had been very well received in Ireland. The question now arose whether to follow it up with an invitation to de Valera and Craig to meet with the government 'to discuss the situation and if possible, reach agreement'. There was some evidence that de Valera would not insist on a republic. The proposal was welcomed by the cabinet, Curzon saying that at each stage he had been in

Policy, pp 75–87.   **59** LGP F 45/9/48 (14 June 1921).   **60** LGP F 45/9/48 (14 June 1921).   **61** Sinn Féin boycotted the elections to both senates. The Senate of Northern Ireland (elected by the House of Commons by PR) was an all-Unionist body with the exception of the Nationalist mayor of Derry, an *ex-officio* member, who refused to attend it. Only the 11 nominees by the lord lieutenant constituted the Southern senate.   **62** Jones, *Whitehall Diary*, iii, pp 78–9.   **63** CAB 21(53) (24 June 1921); Jones, *Whitehall Diary*, iii, pp 79–81.

favour of opening up negotiations with 'the enemy', and he viewed Lloyd George's proposal with the 'intensest satisfaction'.[64] The prime minister was authorized to write to de Valera and Craig.

By the cabinet meeting on 29 June, no reply had been received from de Valera and it was decided to wait before taking further action. Meanwhile, Dublin Castle went ahead with the charade of opening the 'parliament of Southern Ireland'.[65] Sir John Ross, who had just been sworn in as (the last) lord chancellor of Ireland, read a 'Speech from the Throne' to four members from Dublin University and one senator (General Sir Bryan Mahon), and the body adjourned *sine die*.[66]

That British government policy had changed as a result of the king's speech in Belfast is beyond question. Churchill, who approved it at the time, later wrote that 'No British government in modern times has ever appeared to make so sudden and complete a reversal of policy.'[67]

De Valera's reply[68] was sent on 28 June. He asked for time to consult his colleagues and also the representatives of the main minority, the Unionists. He also immediately objected to Craig, the leader of a minority in Ireland, being treated as an equal partner in the negotiations that were being envisaged. That attitude suited Craig, who refused the prime minister's invitation, stating that the negotiations should be between the 26 counties and Great Britain, with Northern Ireland remaining as it was.

The Unionists invited, led by Lord Midleton (but excluding Sir James Craig, who politely refused) met de Valera on 4 July. Midleton suggested a truce as a preliminary to negotiations and offered to raise the question with Lloyd George.[69]

On 5 July Smuts visited Dublin on de Valera's invitation and met the president, with Arthur Griffith and Robert Barton (members of the Dáil cabinet) and Eamonn Duggan, director of Intelligence for the IRA.[70]

Smuts reported back to the Irish Situation committee (reinforced with Birkenhead) on the following day. Generals Macready and Tudor and Sir John Anderson were in attendance.[71] He was not impressed with his Irish hosts.[72] (De Valera was a

---

64 Ironically, de Valera, who had been living with a small staff in a house in Dublin, was arrested by troops on 22 June, and released after a peremptory order from Cope on the following day. After his release de Valera was despondent, felt that his political usefulness was over and was inclined to go to Munster to join the IRA. Lloyd George's letter of 24 June came as a complete surprise. De Valera then decided to stay in Dublin and opened an office in the Mansion House: Longford and O'Neill, *De Valera*, pp 127–9.   **65** *Sturgis Diary*, p. 192.   **66** At a ministerial conference on 24 June, it was solemnly agreed that if only the Trinity four attended the opening of the parliament four days later, it would be 'inadvisable' to elect a Speaker! CAB 53(21) (24 June 1921).   **67** W.S. Churchill, *The World Crisis: The Aftermath* (London, 1929), pp 29ff.   **68** See Dáil Éireann, *Official Correspondence relating to the Peace Negotiations, June-September 1921*, and Cmd 1470, 1502 and 1539 for all the letters between de Valera and Lloyd George.   **69** Longford and O'Neill, *De Valera*, p. 130.   **70** Mansergh points out that Duggan was a prisoner in Mountjoy until the truce, but he could have been released specially for the meeting. Other sources agree that Duggan was present. Mansergh, *Unresolved Question*, p. 163; Longford and O'Neill, *De Valera*, p. 130; *Sturgis Diary*, p. 199.   **71** The fullest accounts of this important meeting are in Jones, *Whitehall Diary*, iii, pp 82–5, and Fisher Diaries (6 July 1921), 17, 103.   **72** The fol-

pure visionary and they were all third rate men, quite incapable of dealing with a big situation.) Smuts found them disposed to refuse the invitation to meet the prime minister[73] – de Valera had not yet replied to the letter of 25 June – mainly because Craig, the leader of a minority party in Ireland, was invited as an equal to president de Valera and the British government would be able to play the old trick of 'divide and rule'.[74] Smuts told them bluntly that to refuse the offer could lead to very grave consequences and destroy whatever public support they had enjoyed.

Smuts believed that he had clearly made an impression on this point. But he was less successful in trying to persuade the Irish to accept dominion status, like South Africa. They insisted that they wanted a free choice between a republic and a dominion, not an imposed solution. They were prepared, if offered a republic, to accept limitations. Smuts could not shake de Valera on this point, and the latter continually harped on the crime of partition, committed by the British government.

At this point Midleton's letter was read, asserting that the Southern Unionists were convinced that a truce would make it impossible to renew the struggle.

The conference then agreed to meet de Valera without Craig and to arrange a truce. General Macready[75] was in favour of an open and formal truce, but Lloyd George preferred a 'gentlemanly understanding' – Balfour, sardonic as always, said: 'A gentlemanly arrangement without gentlemen' – and the conference concurred. On the following day, Macready travelled to Dublin and a truce was agreed. On the same day de Valera, having met the Southern Unionists again, replied to Lloyd George that he was 'ready to meet and discuss with you on what basis such a conference as that proposed can reasonably hope to achieve the object desired'.

In spite of the British desire for an informal truce, de Valera issued a proclamation, regarding the cessation of hostilities (9 July).[76]

The truce came into operation at noon on 11 July and the IRA attacks, which had continued up to the last minute, ceased.[77]

FROM TRUCE TO TREATY

The events between the establishment of the truce in July and the signing of the Treaty in December 1921 constitute the most written-about section of modern Irish history. We will not attempt to reproduce the work of other scholars, but will concentrate on certain key points.

lowing is a conflation of the Jones and Fisher accounts. **73** For the Smuts-de Valera correspondence (July-Aug. 1921) see LGP F 11/3/11–F 45/9/52. De Valera regarded Smuts as 'the cleverest of all the leaders he met in that period, not excluding Lloyd George', Longford and O'Neill, *De Valera*, p. 130. **74** Greenwood assured Craig that the decision to invite him and de Valera was the unanimous opinion of the cabinet and urged him to accept 'in the interests of Ireland and the empire'. Greenwood to Craig (25 June 1921), Craigavon Papers, T 3775/14/3. **75** After Smuts had spoken, Generals Macready and Tudor and Sir John Anderson were brought in. All agreed to a truce. **76** Longford and O'Neill, *De Valera*, p. 132. **77** Over one quarter of all crown casualties during the 'war of independence' occurred between May and July 1921: Boyce, *Englishmen and Irish Troubles*, p. 185.

On 14 July, de Valera met Lloyd George who tried hard to sell him the notion of dominion status, but de Valera resisted the premier's blandishments.

There followed a correspondence between Lloyd George and de Valera (16 letters in all) between July and September 1921. The first letter (20 July) invited Ireland to take her place in 'the great association of free nations', then becoming known as the Commonwealth: dominion status, subject to certain conditions – no navy, a limited army, no customs charges on British goods and responsibility for some of the national debt.[78] De Valera had seen the text of these proposals before he left London and told Lloyd George that he could not recommend them either to the cabinet or the Dáil.[79]

De Valera's considered reply to the letter of 20 July was sent on 10 August. He reiterated that neither the Dáil nor the Irish people could accept the British proposals. The draft was self-contradictory. Ireland's right to self-determination was acknowledged but fettered by conditions. Ireland had an indefeasible right to decide her own destiny. dominion status would be illusory, because of British proximity. The question of the 'political minority' was one for the Irish people to settle. 'We cannot admit the right of the British government to mutilate our country.'[80]

The letter of 10 August contained the two principles on which de Valera refused to yield – the necessity for Irish unity and freedom to choose her own destiny.

Meanwhile, the Second Dáil began its proceedings with a public session on 16 August 1921.[81] The Dáil enhanced the status of de Valera with the title 'President of the Republic'. The president then nominated a slimmed-down cabinet: Griffith (Foreign Affairs), Collins (Finance), Robert Barton (Economic Affairs), Cathal Brugha (Defence), Austin Stack (Home Affairs) and W.T. Cosgrave (Local government). He also appointed seven non-cabinet ministers ranging from Desmond FitzGerald (minister for Publicity) to Count Plunkett, sidelined as minister for Fine Arts and J.J. O'Kelly, better known as 'Sceilg' (rock), (minister for Education), whose main function appears to have been to provide Irish translations of communications with the British government.[82]

In subsequent correspondence Lloyd George failed to shake de Valera's insistence on the issue of independence. The furthest he would go would be a conference on the broad principle of government by the consent of the governed.[83] Tiring of the

78 For the correspondence see Cmd 1470, 1502 and 1539.   **79** Longford and O'Neill, *De Valera*, pp 136–8.   **80** Cmd 1470.   **81** The public session continued on 17 August. The Dáil held several private meetings between 18 August and 14 September, and then adjourned until the Treaty debates. All meetings were held at University College, Dublin, in Earlsfort Terrace.   **82** In the early twentieth century several writers in Irish adopted names in Irish from the world of nature, e.g. Dr Douglas Hyde 'An Craoibhinn Aoibhinn ' (little mellow branch). Probably the last was Pádraig Ó Siochfradha 'An Seabhac' (the hawk), who survived into the 1960s.   **83** At that time de Valera was formulating the doctrine that later became known as external association, viz. that an independent Ireland could be associated with the British Commonwealth without allegiance to the king – the term 'Head of the Commonwealth' was a subsequent gloss. Although approved by the Dáil cabinet, the proposal was not formally raised in negotiations with the British. Long-

apparently interminable correspondence, and faced with de Valera's firmness, Lloyd George called a cabinet meeting for Gairloch near Inverness for 7 September, the purpose of which was to tie the Irish to conditions antecedent to any conference – especially allegiance to the crown. At the improvized cabinet meeting, Lloyd George pointed out that an unconditional invitation would be a sign of weakness. If a break had to come, he would prefer it to be on the issue of allegiance, not on Fermanagh and Tyrone, when at that very time the Nationalist-dominated Tyrone county council had repudiated the authority of the new Northern parliament and Fermanagh council would soon follow.[84]

To Lloyd George's surprise, the cabinet divided equally, three Conservatives joining the Liberals, who set the tone of the discussion. Munro (secretary for Scotland) pointed out that Britain must not forfeit the confidence of the civilized world and that a break would be too awful to contemplate. At a crucial point Andy Cope was brought in. Of all the senior officials in Dublin Castle, he was the one who best understood the republican mentality. Cope told the cabinet 'quite firmly' that if a letter was sent demanding Irish allegiance to the crown, there would be no conference, and if a break occurred, the mass of the Irish people would follow de Valera and not Lloyd George. Cope's intervention appears to have been decisive.[85] It was probably his most important contribution to the discussions on Ireland.

After much consideration, the cabinet argued that the prime minister should write to de Valera in these terms. The principle of government by the consent of the governed is the foundation of British constitutional development, but the government could not accept an interpretation which would commit them to any demands which the Irish leader might present. On the other hand, they invited discussion of the proposals on their merits.[86]

Ministers assumed that de Valera would agree that the correspondence had lasted long enough. They must therefore ask for a definite reply as to whether he was prepared to enter a conference 'to ascertain how the association of Ireland with the community of nations known as the British empire, can best be reconciled with Irish national aspirations'. If so, Lloyd George would suggest the conference would begin at Inverness on 20 September.[87]

De Valera replied promptly.[88] He had no hesitation in accepting the invitation and was summoning the Dáil to ratify the names of the representatives they proposed for the conference on 20 September. There followed an unequivocal restatement of the Irish position: 'Our nation has formally declared its independence and recognizes itself as a sovereign State. It is only as the representatives of that state and as its chosen guardians that we have any authority or powers to act on behalf of our people'.

ford and O'Neill, *De Valera*, pp 139–40.   **84** CAB 23/27, 74(21) (7 Sept. 1921). A more extended account is to be found in Jones, *Whitehall Diary*, iii, pp 106–12.   **85** Jones, *Whitehall Diary*, iii, pp 111–13.   **86** Cmd 1539.   **87** Cmd 1539 (7 Sept. 1921).   **88** Ibid. (12 Sept. 1921).

De Valera knew that that paragraph would infuriate Lloyd George. But as he told a private session of the Dáil on 14 September, the Irish cabinet felt it absolutely necessary to restate their position; otherwise the British press would claim that they had capitulated.[89] He asked the deputies to consider the possible implications of his letter – abandonment of the negotiations and renewal of the war. The deputies unanimously approved of the letter. (In view of his later career, it is interesting that the strongest endorsement of the republican ideal was made by Kevin O'Higgins.) The meeting then went on to consider the appointment of plenipotentiaries for any future negotiations with the British government, as authorized by the cabinet. To the surprise of many, the president did not include his own name. He admitted that the cabinet was divided on this issue, and it was carried only by his own casting vote. His defence was that as the head of state and the 'symbol of the Republic' he would be compromised by any negotiations.

One minister[90] formally moved that the president lead the delegation. ('It was not usual to leave the ablest player in reserve.') The amendment (seconded by Gavan Duffy) was lost. The president, on behalf of the cabinet, then proposed the names of five delegates: Arthur Griffith (chairman), Michael Collins, Robert Barton, Eamonn Duggan and George Gavan Duffy.

Griffith was the obvious chairman. Collins objected to his appointment. He told the Dáil that 'he believed the president should have been part of the delegation. He did not want to go himself and he would very much prefer not to be chosen'.[91] Although de Valera insisted in including Collins, on account of his 'amazing efficiency', he had already experienced doubts about the commitment of Griffith and Collins to the republican ideal.

He also was concerned at Collins' close connection with the IRB, of which he had become president of its supreme council in 1919. After 1916 de Valera (with Brugha) had severed his tenuous connection with that organisation, since he believed a secret society was no longer needed, especially after the election of 1918.[92] Certainly Collins believed he had been chosen as a probable scapegoat if the negotiations failed, but there is no evidence of such a Machiavellian intention on de Valera's part.[93]

To balance Griffith and Collins, de Valera chose 'a staunch republican',[94] Robert Barton, a Protestant landowner from Wicklow, who had been educated at Oxford and served in the British army, but later fully adopted republicanism. He was minister for economic affairs in the Dáil cabinet. The remaining delegates[95] were Eamonn

89 *Dáil Éireann, Private Sessions of Second Dáil* (14 Sept. 1921), pp 87–95. **90** W.T. Cosgrave. **91** *Dáil Debates, Private Session* (14 Sept. 1921), p. 96. **92** Longford and O'Neill, *De Valera*, pp 147–8. **93** Ibid.; Coogan, *Collins*, pp 229–32. By 1921 the IRB, which had masterminded the 1916 Rising and regarded itself as the true revolutionary government, had become 'a rather remote, unreal and shadowy kind of organisation': Ó Corráin (ed.), *Hogan*, p. 219. But Collins maintained his contacts with the Supreme Council and ensured that it approved his nomination as a delegate to London. Leon Ó Broin, *Revolutionary Underground*. **94** Longford and O'Neill, *De Valera*, p. 149. **95** Pakenham, op. cit., pp 130–5.

Duggan, a solicitor, director of Intelligence for the IRA and after the truce liaison officer with the British. He was close to Griffith and Collins. Lastly, the president nominated, and the Dáil approved, George Gavan Duffy, son of the Young Irelander, a solicitor who had defended Roger Casement, former secretary to the Dáil cabinet and envoy to Paris. All five were Dáil deputies, but only three were ministers. Griffith and Collins were by common consent the most important.

At de Valera's personal request, Erskine Childers was appointed secretary to the delegation.[96] Born in England, but brought up in the Wicklow home of his first cousin, Robert Barton, and educated at Cambridge, Childers joined the British civil service and later served with distinction in the navy. Childers was secretary of the ill-fated Irish Convention of 1917, but by 1919 had become a dedicated republican and held several posts under the control of the Dáil, ending as director of publicity (February 1921), and a Dáil deputy from May.[97] The assistant secretary was John Chartres, an Englishman with no Irish connections, a former civil servant, who also embraced Irish republicanism.

The British delegation was dominated by the 'Big Four' – David Lloyd George, prime minister and leader of the Coalition-Liberals, Austen Chamberlain, lord privy seal and Conservative leader, Viscount Birkenhead, Conservative lord chancellor, and Winston Churchill, Liberal colonial secretary.[98] The others were Sir Laming Worthington-Evans, a middle-ranking Conservative minister, who then held the post of secretary for war, Sir Gordon Hewart, a barrister turned Liberal MP, the current attorney-general, and Sir Hamar Greenwood, Irish chief secretary. The Secretaries were men of unusual calibre, Tom Jones the first professor of economics at Queen's University, Belfast, then assistant secretary to the cabinet and Lloyd George's alter ego, and Lionel Curtis, former editor of the *Round Table*, Fellow of All Souls and expert on the Commonwealth.

The credentials issued to the Irish delegates referred to them as 'envoys plenipotentiary from the elected government of the Republic of Ireland', a formula that would not be accepted by the British government.[99] But if the Dáil cabinet regarded the delegates as plenipotentiaries, then, according to the customary usage, they should have full powers to negotiate and sign a treaty, like the delegates at Versailles. Nevertheless, President de Valera issued a written instruction (in the name of the cabinet) directing that 'the complete text of the draft treaty about to be signed' be submitted to the cabinet in Dublin and a reply awaited. These contradictory instructions led to later confusion and dissension.

De Valera's letter of 12 September was the last letter in the correspondence. It was followed by an exchange of telegrams (eight in all) between the two leaders.

96 For a sympathetic picture of both delegations see Pakenham, op. cit., pp 123–43. **97** Childers was best known in Britain and Ireland for his pre-war novel, *The Riddle of the Sands*. **98** Interestingly, Pakenham refers to Birkenhead as 'the greatest of all (four)', op. cit., p. 123. **99** Nor were the credentials formally presented. Mansergh, *The Unresolved Question*, pp 176–7.

On 15 September Lloyd George repeated what he had told two emissaries sent by de Valera, that the reiteration of the Irish claim to negotiate with the British as the representatives of a sovereign state would make such a conference impossible. De Valera feigned 'surprise' that Lloyd George did not see that for the Irish to go into a conference without making their position equally clear would lead to misunderstandings and would irreparably prejudice their cause (16 September). Lloyd George then pointed to the international consequences that would follow a 'formal and official recognition of Ireland's severance from the King's domains'. A conference would be impossible while the Irish insisted on such a claim (19 September).

De Valera then sent the disarming reply that in accepting the invitation conveyed in the letter of 7 September, 'we have not asked you to abandon any principle, even informally, but surely you must understand that we can only recognize ourselves for what we are' (17 September). After two further telegrams, Lloyd George relented, and while repeating that the government's position was 'fundamental to the existence of the British empire', issued a fresh invitation to a conference in London on 11 October, 'with a view to ascertaining how the association of Ireland with the community of nations known as the British empire may best be reconciled with Irish national aspirations'. On the following day (30 September) De Valera telegraphed his acceptance.

While the Treaty delegates have frequently been accused of allowing themselves to be mesmerized by the 'Welsh wizard', it must be noted that in the exchanges described above de Valera showed himself to be a skilful negotiator. He did not concede any argument advanced by Lloyd George and maintained the Irish claim to independence throughout.

The negotiations that led to the Anglo-Irish Treaty of 6 December 1921 fall into three phases. In the first (11 October to 3 November), after preliminary exchanges, the Irish delegates agreed to concessions to the British government, on condition that the 'essential unity' of Ireland was maintained, thus indicating the centrality of the Ulster issue. The second phase lasted from 5 to 16 November, during which Lloyd George tried hard to persuade Sir James Craig to join the conference. By 16 November the British realized that if there was to be any agreement, it would have to be between Great Britain and 'Southern Ireland', and determined to persuade the Irish delegates to accept the British proposals – to which they agreed on 6 December.

Phase I (11 October – 3 November)
This phase covers seven plenary sessions and three 'sub-conferences' – 30 October and 1–3 November.

The Irish delegates had been authorized by the cabinet to present to the British 'Draft Treaty A'[1] – the main points of which were sovereignty for the Irish state and external association (de Valera's brainchild) – Ireland to be associated with the Com-

---

1 For the text of Draft Treaty A see *Documents on Irish Foreign Policy* (henceforth *DIFP*), i, *1919–1922* (Dublin, 1998), (7 Oct. 1921), pp 271–2.

monwealth for matters of common concern, but without membership.[2] However, these proposals were not incorporated into a document,[3] so the plenary sessions began with a discussion of the Irish objections to the proposals of 20 July.

At the first sessions, Griffith went on the attack. The British proposals, he claimed, would result in their dominating Ireland, both politically and militarily. Lloyd George disclaimed any intention to dominate Ireland militarily. All the British wanted was free access to the Irish coasts in the case of war. Then, when Griffith and Collins argued for Irish neutrality in the event of a war declared by Great Britain, the British delegates flatly refused to consider the question. 'To repudiate British wars means leaving the empire', said Lloyd George, denying that a dominion could remain neutral in a war involving Britain.

The question of trade also came up. Lloyd George, a lifelong free trader, did not desire any economic barriers between the two countries, but Collins and Barton presented a strong case for Irish protection for the growth of Irish industries and against dumping of cheap British goods.[4] The financial question – Ireland's responsibility for a share of British national debt, as against the Irish claim of over-taxation during the century of the Union – was referred to a sub-committee.

So far the conference had gone well! There was a general feeling that differences on these issues could be overcome, but the first of two crucial problems, that of Ulster, came up at the fourth plenary session on 14 October, two days before the delegates received the 'Ulster clause' (not included in Draft Treaty A) from the cabinet in Dublin. Griffith and Collins set out to disabuse the British of the notion that the six counties were a homogeneous unit and rehearsed the arguments that were to become familiar in later decades. Partition under the 1920 Act was unnatural. It included in Northern Ireland two counties with Catholic majorities. In the whole of the six counties one third of the population was Catholic. Collins claimed that if the British did not intervene, nationalist Ireland could come to an agreement with Ulster.

---

2 For the details of the Treaty negotiations, the cabinet papers are not very useful, since the first cabinet meeting on Ireland after 11 October was on 10 November. Lloyd George preferred to work through a few ministers and his secretariat. *Peace by Ordeal* (London, 1935) by Frank Pakenham (later the earl of Longford) is the first, and in many respects the best, detailed study. Its defect is that the author did not have access to British unpublished sources. (British cabinet papers were not available to scholars until the 1960s). The *Whitehall Diary* of Tom Jones, assistant secretary to the cabinet, especially iii, pp 119–85, complements Pakenham's account. Sheila Lawlor's useful work, *Britain and Ireland, 1914–23* (Dublin, 1983), largely based on unpublished sources, relies heavily on the Jones diary in chapter 5 (October-December 1921), pp 113–46. The Griffith-de Valera correspondence during the negotiations is reprinted in *DIFP*, i, *1919–1922*, pp 274–360.
3 In a memorandum John Chartres provided a neat exposition of external association. The Irish desired a republic, he wrote, but the British would not sacrifice the king. Ireland could not associate herself with the Commonwealth without reference to the monarch, so as 'the germs of a compromize' he proposed that a sovereign Ireland would recognize the king as head of the Commonwealth, but that he would have no role in Ireland's internal affairs. Chartres to Griffith (14 October 1921), *DIFP*, i, pp 276–7. See also Pakenham, op. cit., pp 241–3, 376–7. **4** By 1921 the dominions had the right to impose tariffs.

The same argument was advanced by Griffith on 17 October. The 'Ulster clause' had arrived from Dublin, but Griffith did not present it.[5] The clause prescribed that the six counties would be represented directly in a Dublin parliament, or if they, or any substantial part, refused the offer, they could keep the parliament already established in Belfast, but subject to the Dublin parliament, not Westminster as under the 1920 Act. The great weakness of the Ulster clause was a lack of any provision to deal with the anticipated rejection by the Craig government, so Griffith did not present it to the conference, but argued with a wealth of detail that if the British stood aside, Sinn Féin could make a 'fair proposal' to Ulster. ('We are prepared to reason with them and perhaps obviate a vote; if they refuse, to give free choice to the people in the area.' But he did not prescribe how the free choice was to be exercised.[6]) The discussions on Ulster were inconclusive.

The second crucial issue, allegiance to the crown, was precipitated into the negotiations by external events. Pope Benedict XV sent a telegram to king George, rejoicing at the resumption of the Anglo-Irish negotiations and praying for their success. The king replied with thanks, praying that the conference might achieve a permanent settlement of the troubles in Ireland. This exchange prompted de Valera to send a telegram to the pope, pointing to the 'ambiguities' in the king's reply and asserting that the Irish people did not owe allegiance to the British king: 'The independence of Ireland has been formally proclaimed ... The trouble is between England and Ireland.'[7]

British politicians reacted unfavourably to what they considered an insult to both pope and king. At the sixth plenary session after a report from a sub-committee rejecting the Irish claim to neutrality as incompatible with British security, Lloyd George challenged Griffith to provide 'clear and definite' answers to three questions: 1) allegiance to the crown, 2) membership of the Commonwealth, and 3) naval facilities.

'Draft Treaty A' was presented to the conference, too late for full consideration by the British at the next plenary session on 24 October.[8] At the meeting Lloyd George questioned the definition of external association 'for all purposes of agreed common concern'. Did that mean Ireland within the empire? During the discussion Griffith made his first concession – recognition of the king as head of the British empire, with which Ireland would be associated, with reciprocal citizenship rights. On the same evening the first 'sub-conference' took place – a meeting arranged by Cope between Griffith and Collins, Lloyd George and Chamberlain, Griffith insisting that Irish recognition of the crown, in any form, required the 'essential unity' of the country.

When de Valera received the minutes of the sixth plenary session from Childers (as he had received accounts of earlier sessions) he wrote to Griffith, questioning

---

5 See *DIFP*, i, p. 277 (de Valera to Griffith, 14 October 1921). 6 Griffith suggested constituencies or Poor Law areas as the appropriate units, but Birkenhead said a constituency basis was 'not practical': Jones, op. cit., pp 134–7. 7 Pakenham, op. cit., pp 165–6. 8 For the text see *DIFP*, i (7 Oct. 1921), pp 271–2.

recognition of the crown as head of the association, since there was no mention of the crown in the Ulster clause. This message put the delegates very much on their dignity. A letter, signed by all, protested at this intrusion into the rights of the plenipotentiaries to free discussion. De Valera withdrew his charge and admitted to a 'misunderstanding'.

The seventh plenary session was the last. The sub-conference format appealed to Lloyd George, since he wished to exclude Worthington-Evans and Greenwood on his side and Childers, whom he regarded as a fanatic, on the Irish side. Sub-conferences were held on 26 and 30 October and 2 and 3 November.

The British delegation appeared to have been surprised and pleased by the conciliatory tone of the sessions on 24 October and felt that real progress had been made. At the second sub-conference on 25 October,[9] Griffith and Collins met Chamberlain and Hewart to discuss Ulster. The discussion was inconclusive, but the Irish were left with the impression that a possible solution might be for Northern Ireland to keep its existing parliament, but subordinate to a Dublin parliament. However, at a private meeting of the British delegates, Lloyd George outlined the immediate strategy – to shelve the Ulster issue ('If they accept all subject to unity, we are in a position to go to Craig; if they don't, the break is not on Ulster').[10] They decided to ask the Irish for their views in writing on the issues of allegiance, the empire and defence.

The Irish document did not appear until 29 October. If confirmed the concession already made, that Britain could have the coastal facilities required, but stood firm on the other points. There was no mention of allegiance to the crown or common citizenship, but a recommendation that a free and undivided Ireland might recognise the crown as the accepted head of the Commonwealth.

Through the assistant secretary to the cabinet, Tom Jones, Lloyd George and Chamberlain informed the Irish delegates that their document was 'most unsatisfactory'. It was not published, but at a meeting on 30 October between Griffith and Lloyd George, the latter asked for an assurance on the main points (allegiance, the empire and defence), which could be used on the following day as a defence against a censure motion by 'diehard' Conservatives in the House of Commons. The prime minister also asked for a letter that might be presented to the annual Conservative Party conference on 17 November – a more difficult hurdle to overcome, since Lloyd George as a Liberal could not attend, but hoped that Chamberlain and Birkenhead might carry the day for the Coalition. Griffith gave a verbal assurance that, provided he was satisfied on the other points, especially the essential unity of Ireland, his delegation would recommend recognition of the crown. The other matters were left over.[11]

Armed with this assurance, Lloyd George crushed the Diehard opposition in the Commons on 31 October by 437 votes to 43. Meanwhile, Griffith prepared a letter

9 Jones, op. cit., pp 145–6. Pakenham gives the date as 26 October, op. cit., p. 190.   **10** Jones, op. cit., p. 146.   **11** Later at this meeting, which 'some would say (was) the most important in

(which he intended to be a personal assurance), to be presented to the Conservative conference.

Recapitulating the conversation of 30 October, Griffith assured Lloyd George that, provided he was satisfied on every point, he was prepared to recommend recognition of the crown and 'free partnership' with the British Commonwealth, the form of both to be decided later.

When the other delegates were informed, Barton and Gavan Duffy were furious,[12] firstly at the suggestion of a personal letter, also at its terms. Gavan Duffy strongly advised Griffith not to send the letter, since its main effect would be 'to undermine the stand we have taken'.[13]

Faced with this opposition, Griffith gave way[14] and agreed to send a letter from the entire delegation, despatched on 2 November, with the significant alteration 'recognition of the crown as head of the proposed Association of Free States'. At a meeting between Lloyd George, Chamberlain, Birkenhead, Griffith and Collins on the same day, Birkenhead persuaded the Irish to accept a further change: 'free partnership with the other states associated within the British Commonwealth'.[15]

Forwarding de Valera a copy of this letter,[16] Griffith explained that the British had assured him that if Ulster proved unreasonable, they were prepared to resign 'rather than use force against us'.

Phase II (3 November – 16 November)
Lloyd George did indeed tell his closest friends that if Ulster remained obdurate, he would resign, but his sincerity may be doubted, since he had made a similar promise in 1916 and subsequently reneged on it. He also threatened to call an election if he met with 'unreasonable' opposition.[17]

On the strength of the assurances from the Irish delegates, Lloyd George sent to Craig on 10 November a carefully drafted letter,[18] stating that the time had come for formal consultations between London and Belfast, since the settlement which the government believed 'is not attainable' would comprise the following: 1) Ireland's allegiance to the throne and membership of the Commonwealth, 2) naval facilities indispensable for British security, 3) the government of Northern Ireland to retain

the whole negotiations', Pakenham, op. cit., p. 193. Collins and Birkenhead joined the discussion. **12** Childers, though not a delegate, agreed with Barton and Gavan Duffy. Between 20 October and 3 November he sent to Griffith several memoranda, explaining in meticulous detail why Ireland could not accept dominion status, *DIFP*, i (20 Oct. 1921), p. 300*ff.*   **13** Pakenham, op. cit., p. 196.   **14** This was the first instance of a division within the delegation between Barton and Gavan Duffy and the others. The split was to widen over the succeeding weeks. Griffith to de Valera (3 Nov. 1921), *DIFP*, i, p. 300.   **15** LGP (2 Nov. 1921), F/21/1/1.   **16** Griffith to de Valera (3 Nov. 1921), *DIFP*, i, p. 300.   **17** Lord Riddell, *Intimate Diary of the Peace Conference and After* (London, 1933), pp 330–2 and Scott, *Diaries* (28–29 Oct. 1921), pp 402–5. Riddell and Scott were newspaper proprietors friendly to Lloyd George.   **18** The letter was largely crafted by Lionel Curtis and Sir Edward Grigg (later Lord Altrincham), private secretary to the prime minister. See Deborah Lavin, *From Empire to International Commonwealth: A Biography of Lionel Curtis* (Oxford, 1995), pp 190–1.

the powers conferred by the 1920 Act, and 4) an all-Ireland parliament, leading to a self-governing Irish state.

The letter went on to admit that Northern Ireland might fear that the all-Ireland parliament might discriminate on matters of patronage, education and trade, and invited Craig and his cabinet to discuss the means of dealing with appointments, revenue collection and safeguards concerning imports and exports. Income tax, customs and excise were reserved to Westminster by the 1920 Act, but Lloyd George thought that 'grave difficulties' would arise if the reserved services were not conferred on a common authority.[19]

As a sweetener, the letter pointed out that if Northern Ireland became part of an Irish dominion, its contribution to imperial charges would be voluntary. Hinted, but not stated, was the inducement that an Irish dominion could settle its own income tax at a lower rate than that of the United Kingdom. On the other hand, if Southern Ireland became a dominion and the North remained as it was with representation in the imperial parliament, it was clear that the people of the area would have 'to bear their share of all Imperial burdens', in common with other British taxpayers.[20]

The letter ended by pointing out that Great Britain, for the sake of a settlement, was willing to forego any contribution from Ireland to imperial expenses and 'cordially' invited the Northern ministers to a conference with the British government.

The question of the area within the special jurisdiction of the Northern parliament was reserved for further discussion.

Craig, who was in London at the time, replied from the Constitutional Club[21] on the following day. Behind the customary protestations of loyalty to the crown and empire lay a flat refusal. An all-Ireland parliament was 'precisely what Ulster has for many years resisted with all the means at her disposal', and her detestation was as strong as ever. Moreover, the area had been very carefully considered when the 1920 bill was going through parliament, and the question should not be reopened. Any further discussion would be fruitless, unless the proposal for an all-Ireland parliament was withdrawn.

Craig also suggested that the reserved powers under the 1920 Act be transferred to both governments – in other words, that two dominions be set up in Ireland.

Lloyd George's reply (14 November) expressed a feeling of resentment that a preliminary limitation on freedom of discussion between two sets of ministers of the crown should have been made. The proposal for two dominions in Ireland was 'inde-

19 See *Correspondence between His Majesty's Government and the Prime Minister of Northern Ireland relating to the Proposals for an Irish Settlement* (Cmd 1561) for the letters between Lloyd George and Craig (10, 11, 14, 17, 18, 20 November 1921). They are reproduced in Appendix 2. **20** Lloyd George told the cabinet meeting on 10 November that 'if Ulster wishes to pay 12/- in the £, her position is strong. She can't expect to get all the benefits of both systems', Jones, op. cit., p. 161. See also pp 157–62. CAB 21 (8) (10 Nov. 1921) briefly records that the cabinet gave their general approval to the policy followed by the British representatives and expressed satisfaction with the results so far. **21** As a former junior minister in London, Craig was well known in this bastion of Conservatism.

fensible'. Three national governments in the British Isles would be 'ruinous' to trade, not to speak of the welfare of minorities.

In subsequent correspondence, Craig hinted that the dominion proposal was in response to the obvious intention of Lloyd George's letters – the exclusion of Northern Ireland from the United Kingdom. Further letters showed no change in Craig's attitude.

These exchanges, the most important during the Anglo-Irish negotiations since they determined the future course of Irish history, have been regarded by Irish general historians as a doomed venture by Lloyd George. 'Sir James Craig had in fact only to sit tight',[22] Lyons writes, while Foster states that 'Lloyd George's attempts to bring Craig into negotiations, and adapt Northern Ireland's 1920 status, met with a predictably monolithic response',[23] and Lee writes, 'Lloyd George could not, even if he wanted to, deliver unity in the face of resistance by Ulster Unionists, supported by the Conservatives.'[24] But was Craig's position so secure? To answer that, we must go back a little.

Lloyd George had been fully aware of the attitude of the Belfast cabinet since a special meeting of the full cabinet at 10 Downing Street with Lloyd George in the chair on 18 July.[25] Lloyd George asserted that in negotiations with de Valera he hoped to persuade him to give way on allegiance, naval facilities and free trade. On the question of Northern Ireland de Valera was willing to accept the situation under the 1920 Act, but only provided that the Belfast government accepted an all-Ireland parliament.

Lloyd George invited the cabinet to put forward proposals 'to meet the case', but the Northern ministers refused to co-operate. Craig considered that Northern Ireland had gone as far as it could, and the only way out was to give 'every possible concession' to Southern Ireland and leave Northern Ireland alone. He refused Lloyd George's invitation to a further meeting, since 'it would serve no useful purpose'.[26]

That uncompromising attitude was reinforced by Craig's public refusal to join the negotiations, but within three months the situation had changed. On the one hand, there were riots in Belfast, which the government was unable (or unwilling) to control, while the county councils of Tyrone and Fermanagh and Derry city council were threatening secession. On the other hand, parliamentary and public opinion in Great Britain was becoming favourable to an Irish settlement. The House of Commons, elected in December 1918, was different from its predecessor of eight years previously. The Liberals (reduced from 272 to 161) and Labour (increased from 42 to 83)

22 Lyons, *Ireland since the Famine*, p. 433.   23 Foster, *Modern Ireland, 1600–1972*, p. 506.   24 Lee, *Ireland, 1912–1985*, p. 52.   25 CAB (NI) 4/3 (18 July 1921). The minute is dated '18 June', but this must be a misprint, since Lloyd George first met de Valera on 14 July.   26 On 28 November and 16 December respectively, Tyrone and Fermanagh county councils withdrew recognition from the Belfast government and placed themselves under the Dáil. Several smaller authorities followed suit. The government's reaction was to abolish these authorities and in the following year to introduce legislation abolishing PR for local elections: S. Elliott, 'Electoral Systems', pp 214ff.

were still (as formerly) strongly for an Irish settlement, but the greatest change occurred in the ranks of the Conservatives, up from 272 to 335.[27] Whereas the pre-war Conservatives were greatly concerned with imperial and constitutional issues, their successors were more involved with socio-economic problems in Britain and wished to be rid of the Irish problem.

Another difference with 1914 was that, as Boyce concludes, 'almost without exceptions, the leading organs of national and provincial Conservative opinion urged the Ulstermen to play their part in a general settlement'.[28]

To test opinion within the party, Chamberlain sent identical letters to the chief whip, Sir George Younger, and the party chairman, Sir Leslie Wilson, posing the question, if Sinn Féin made all the concessions the British deemed essential and the negotiations broke down because Ulster refused to join an all-Ireland parliament, 'Could we rally the country to the support of Ulster in this attitude and for the coercion of Southern Ireland on this basis?' Both replied in the negative, Wilson asserting that the then House of Commons was very different from its predecessors.[29]

Armed with these reports, Chamberlain replied in a public letter of 14 November to an equally public letter from Ronald McNeill (an Ulster-born Conservative MP at Westminster) challenging him to make public his attitude to an all-Ireland parliament. While disclaiming any intention to coerce Ulster, Chamberlain insisted that that promise was not incompatible with seeking peace ('our greatest need') between Ireland and Great Britain. One and three quarter million people were unemployed, trade was stagnant and confidence lacking, but these ills could be coped with if the existing negotiations succeeded, on the result of which depended 'perhaps the future peace of the world.'[30]

In the same letter, Chamberlain indicated that he was satisfied with the 'plain' assurances from Sinn Féin that if 'all other points are successfully arranged', they would recommend the acceptance of the essential British conditions, and added: 'We are therefore face to face with the Ulster problem.'[31]

With the same idea in mind Lloyd George invited Craig to Downing Street on 5 November. While Lloyd George's account of the meeting to the cabinet on 10 November that he found Craig 'reasonable' and willing to discuss the conditions under which an all-Ireland state might function[32] might be regarded as self-serving, it is backed

27 This figure refers to Coalition-Conservatives and excludes the 23 Conservatives who did not support the Coalition and the 26 Irish Unionists. The Representation of the People Act 1917 nearly trebled the electorate, from 7.7 million in December 1910 to 21.4 million in December 1918. 28 Boyce, *Englishmen*, p. 161. In 1921, *The Times* (briefly) and the *Daily Mail* were owned by Lord Northcliffe, and the *Daily Express* by Lord Beaverbrook, both strong advocates of an Anglo-Irish settlement.   29 Chamberlain to Younger and Wilson (8 Nov. 1921), Younger and Wilson to Chamberlain (9 November 1921), ACP 31/3/40, 41–2; WEP, 910m f194–7.   30 *The Times* (15 Nov. 1921). In an earlier letter to his wife, Chamberlain emphasized his commitment to an Irish settlement. He was aware that Bonar Law, recovered from his illness, was waiting in the wings; but wrote scornfully that 'his courage is not equal to his ambition'. A miscalculation. Austen to Ivy Chamberlain (4 Nov. 1921), ACP, 6/1/449.   31 Ibid.   32 This account is entirely based on

by two independent pieces of evidence: H.A.L Fisher wrote in his diary (10 November) that Craig was 'at first very pliant', and Frances Stevenson wrote that by the evening of 5 November Lloyd George 'had extorted from (Craig) considerable concessions, the most important being an all-Irish parliament, which we believe will satisfy the Sinn Féiners'.[33]

5 November 1921 was a Saturday. On Sunday Craig called on Bonar Law. There is no record of the meeting, 'but it is not difficult to surmise the tenor of the conversation'.[34] In a long letter to an old friend, J.P. Croal (editor of the *Scotsman*)[35] at this very time Bonar Law summarized his political credo. There were, in his long career, only two things he regarded as matters of conviction – tariff reform and fair play for Ulster – 'the rest was mainly a game'. Such niceties as Tyrone and Fermanagh did not trouble Bonar Law. Territorial Ulster was whatever the Unionists determined it to be. If Lloyd George persisted in his plan to 'bully' Ulster, Bonar Law would oppose him and, if necessary, split the Conservative Party in the attempt.[36]

Further confirmation of Bonar Law's role in influencing Craig is to be found in Craig's speech to the Belfast House of Commons on 28 November. Having lambasted the British press for its 'skilful, malignant and unrelenting' campaign 'against the people of Ulster',[37] he went on to acknowledge 'the great encouragement we all received by the open, frank, courageous and splendid attitude of Mr Bonar Law'. (In his report of meetings with Lloyd George Craig asserted that his attitude did not change between 5 and 8 November.)

Craig's response was ominous. If the Belfast government would not voluntarily enter, or even discuss, an all-Ireland parliament, and coercion was out of the question for a Conservative-dominated British government, then a major part of British strategy for an Irish settlement would be in ruins.

Tom Jones recorded that after the second meeting with Craig Lloyd George was 'more depressed than I had seen him at all since the negotiations began'.[38] He asked Jones to alert Griffith and Collins to the danger of the negotiations breaking down.[39] Though generally pessimistic, the prime minister saw just one possible way out and asked Jones to put it to Griffith and Collins on the following day: dominion status for the South, the status quo for the North and a Boundary commission to delimit the Northern area if the Unionists remained obdurate.

When Griffith and Collins considered the offer of a boundary commission (8 November), they reacted differently. Griffith was not unduly worried. As he wrote

Jones, op. cit., p. 160. The cabinet minutes are quite uninformative. **33** *Fisher Diaries* (10 Nov. 1921), p. 240; A.J.P. Taylor (ed.), *Lloyd George: A Diary by Frances Stevenson* (London, 1971), (6 Nov.), p. 231. **34** Adams, *Bonar Law*, p. 303. **35** Bonar Law to Croal (12 Nov. 1921), BLP, 107/1/83. **36** BLP, 107/1/93. **37** *Northern Ireland Debates: House of Commons*, i (29 Nov. 1921), pp 288–302. This part of the speech is cited in Boyce, *Englishmen*, p. 162. **38** Curzon told the cabinet on 10 November that Craig had called on him two days previously in a mood of 'extreme disappointment'. He spoke of being 'betrayed, surprised, dismayed, turned out of the British system': Jones, op. cit., p. 161. **39** Ibid., p. 154.

to de Valera, such a commission acting according to the wishes of the inhabitants would be likely to give 'most of Tyrone, Fermanagh and part of Armagh, Down etc.' to the Southern state.[40] He also (unrealistically) suggested that the Northerners were engaged in a 'gigantic piece of bluff'. Collins was more perceptive, opposing the idea of a commission, because if would sacrifice unity entirely. On the following day Griffith told Jones that although the Boundary Commission was the idea of the prime minister, not the Irish, they would not queer his pitch, although they would prefer a plebiscite.[41]

But a mere verbal assurance would be insufficient for Lloyd George. Jones kept warning the Irish of how precarious Lloyd George's position was and of the danger of a hardline Conservative government (under Bonar Law) coming to power. It was highly desirable that the Irish give a written assurance that they would not oppose the Ulster proposals, which might appease the Conservative conference on 17 November.

On 12 November, at a meeting with Lloyd George (at which he was apprised of the first letters to and from Craig), Griffith was sufficiently convinced of the importance of supporting the Coalition during the Liverpool meeting to give an assurance that the Irish would not obstruct the Boundary commission proposal.[42] On the following day Jones drafted a memorandum, to which Griffith agreed. The memorandum[43] stated that if Ulster did not accept the principle of an all-Ireland parliament, she would continue to retain all existing rights under her own parliament with representation at Westminster and British taxation, except as modified by the 1920 Act. In this case, however, it would be necessary to revise the boundary of Northern Ireland according to the wishes of the inhabitants by a Boundary commission.

This was the famous 'letter' that Lloyd George would use to such devastating effect on 5 December. Much has been written as to Griffith's motivation: that he failed to realize that what he regarded as a tactical manoeuvre to help the Coalition at the Liverpool conference, would be interpreted as a definite pledge. Several writers argue that Griffith, a person of unimpeachable integrity, was mesmerized by the 'Welsh wizard', and also that he was foolish in not informing his fellow delegates or the cabinet in Dublin. After an extensive review of the secondary literature, Maye concludes that 'the real beneficiaries of these developments were the Northern government. They had called Lloyd George's bluff, which Griffith had singularly failed to do'.[44]

Living in London, it is not surprising that the Irish delegation were subject to British pressures, especially from their persistent interlocutor, Tom Jones. But de

---

40 Griffith to de Valera (8 Nov. 1921), *DIFP*, i 305.   **41** Jones, op. cit., p. 156.   **42** B. Maye, *Griffith* (Dublin, 1997), pp 202–4. Winston Churchill also expressed fears that the government might abdicate and be replaced by a reactionary Conservative government under Bonar Law, which would be 'a very great public disaster': LGP F/10/1/40 (Churchill to Lloyd George), (9 Nov. 1921).   **43** See memorandum reprinted in R.C. Barton's notes on the sub-conference at 10 Downing Street (5 Dec. 1921), *DIFP*, i, 352–3. It appears that Griffith had seen the memorandum but had not actually signed it.   **44** Maye, *Griffith*, p. 211.

Valera in Dublin was free from such pressures. It is therefore astonishing, in view of his oft-repeated assertions about Irish unity, that de Valera does not appear to have adverted to the implications of the Craig-Lloyd George exchanges mentioned above, although he was certainly informed of them in Griffith's frequent letters.[45]

At the Liverpool conference on 17 December, the government won easily. Chamberlain and Lords Derby and Birkenhead travelled to defend the Anglo-Irish negotiations, and Birkenhead managed to secure the support of Sir Archibald Salvidge, the Tory boss in Liverpool.[46]

At the conference, the leading dissident, Colonel Gretton, proposed a resolution, calling for 'the condemnation of the long-continued ascendancy of crime and rebellion in Ireland', which, if carried, would have forced the Conservative ministers to leave the government; but Salvidge proposed an amendment, defending the Anglo-Irish negotiations, which was carried by 1730 votes to 79. The Tory opposition was defeated, at least temporarily.[47]

However, by 16 November Lloyd George had given up hope of including Ulster in the settlement. He was determined to reach an agreement[48] and in the succeeding week would work on the Southern delegates. The second phase was over.

Phase III (16 November – 6 December)

On 14 November the prime minister resolved to present proposals for a treaty to the Irish delegates within three days. On 15 November he told Griffith and Collins to expect a 'Draft Treaty' on the following day, and Griffith duly informed de Valera. But what was handed to them on that day (by Tom Jones) was not a draft treaty, but 'tentative suggestions' for one.[49]

Essentially, the tentative suggestions were the same as the proposals in the British letter of 20 July:[50] Ireland to become a dominion (like Canada, it was specified); naval defence to be provided exclusively by the United Kingdom, until an arrangement was reached between the two governments for the coastal defence of Ireland; such Irish harbour facilities as the British government might require; no customs duties between the two countries and Ireland to contribute both to the public debt of the United Kingdom and to war pensions. But there was an additional provision. Northern Ireland would be given between six and twelve months to opt out of the

---

45 There is no mention of the correspondence either in Longford and O'Neill, *De Valera*, or John Bowman, *De Valera and the Ulster Question, 1917–1972* (Dublin, 1989).   **46** A good example of Birkenhead's aggressive method of persuasion is to be found in the account of the opening of his (secret) meeting with Salvidge: 'Give me 20 minutes. Don't interrupt me. Don't argue. Don't raise any point till I have finished': S. Salvidge, *Salvidge of Liverpool* (London, 1934), p. 202.   **47** At the conference, General Prescott-Decie, a Southern Irish Unionist of outspokenly far-right and anti-Semitic views, asked for '50,000 additional troops and a free hand' to crush the 'murder Junta'. He was followed by Lord Midleton, who, paraphrasing Tacitus, told the conference that such a policy could make Ireland a desert, but not a loyal, contented and prosperous land. He believed the majority of Southern Unionists desired a settlement. Salvidge, op. cit., pp 205–13.   **48** *Stevenson Diary* (14 Nov. 1921), p. 237.   **49** *DIFP*, i (16 Nov. 1921), pp 309–11.   **50** See above, p. 122.

new dominion by a joint address of both houses of the Belfast parliament. If that were to happen, a boundary commission would be set up to adjust the boundary between the two states 'in accordance with the wishes of the inhabitants', and Northern Ireland would remain part of the United Kingdom.[51]

When sending the proposals to de Valera, Griffith advised that the Irish should reply in a similar fashion – 'tentatively' – but warned that the 'crucial question of crown and empire' must be dealt with in the following week. ('If 'Ulster' gets us to break on them, she will have won the game!')[52]

On 22 November the Irish delegation presented Jones with a memorandum in reply to the British proposals.[53] In many respects it represented an advance on their original instructions of 7 October. Instead of a claim for recognition of Ireland as a sovereign and independent state, there was a declaration that all legislative and executive authority would be derived from the elected representatives of the Irish people; external association with the Commonwealth was firmed up by recognition of the crown as the symbol and head of the Association. They played for time on naval facilities by asking for these to be precisely defined, and an arbitration tribunal was requested to deal with competing claims by Britain for payments towards the war debt and pensions and by the Irish for over-payment of taxes during the Union. The Irish made a strong case for fiscal autonomy, coupled with free trade between the two countries. The whole memo was predicated on the assumption that the 'essential unity' of Ireland was maintained.[54]

Childers was the author of the Irish reply. He also prepared for internal circulation a memorandum going over the British proposals point by point, and concluding that of the ten paragraphs only numbers one (authority derived from the people) and seven (arbitration and debts) did not compromise the claim to complete republican status. This only deepened Griffith's aversion to Childers, whom he regarded as a republican fanatic, capable of wrecking the settlement which Griffith so desired.

However, Griffith was to encounter more formidable opposition, when Lloyd George reacted to the Irish proposals with the same petulance with which he had greeted de Valera's letters some months previously.[55] The document filled him with despair, he told Tom Jones. 'The Irish were back on their independent state again.' Naval demands were not met and there were no safeguards for Ulster. ministers were busy men and, though they had spent many weeks on this matter, had made no progress. He instructed Jones to ask the Irish whether they were to be in or out of the empire.

Tom Jones (who thought the proposals better than previous documents) on meeting Griffith and Collins, did his best to be conciliatory, while apprising them of the danger of a breakdown.[56] Griffith pointed out that the Irish were prepared to give

51 *DIFP*, i, pp 309–11. **52** *DIFP*, i (18 Nov. 1921), p. 311. **53** *DIFP*, (22 Nov. 1921) i, pp 311–13. **54** *DIFP*, (22 Nov. 1921) i, pp 311–12. **55** Jones, op. cit., (22 Nov. 1921), p.170. **56** Jones, op. cit., pp 171–2.

Ulster all the safeguards she desired, but wished her to name them first. (He did not realize that Lloyd George meant safeguards for a *separate* Ulster.) On the other points, the navy and free trade, he indicated a willingness to compromise, but he was not prepared to give the British a blank cheque. On association with the Commonwealth Griffith was firm. Lloyd George threatened that if the document were not withdrawn he would immediately end the negotiations.[57]

At the next meeting on 23 November, a break seemed less likely. Some misunderstandings were cleared up – the British demand for naval facilities was for a war situation only. On this occasion John Chartres attended as a constitutional lawyer and Hewart, the British attorney-general, made his only appearance at the sub-conferences. Chartres' account (on which Pakenham relied) indicates that he expounded the various Irish objections to the crown,[58] but Jones grants him a less exalted role. 'Chartres started some historical disquisition and was shut up brusquely by F.E. [Birkenhead]'.[59]

The Irish then agreed to consult Dublin as to the limits of their association with the Commonwealth. The delegates travelled to Dublin for a cabinet meeting on 25 November, which unanimously approved a resolution that Ireland should recognize the British crown as the symbol and accepted head of the combination of associated states.[60]

When the Irish returned to London the gap between external association and dominion status seemed still unbridgeable. At a meeting with Griffith, Collins and Duggan on 28 November, Lloyd George sought to allay fears of future British interference in Irish internal affairs by offering to insert in the Treaty any phrase ensuring that in Ireland the crown should have no more authority than in Canada, or the other dominions.

The Irish were 'well pleased' with this development, but on the following day a speech by Craig should have blown sky-high any realistic hope of doing a deal with the Belfast government. Addressing the Belfast House of Commons, Craig repudiated, in the strongest terms, the 'preposterous' proposal of an All-Ireland parliament.[61] 'I do not believe that there is a single Member of this House or a single man in Ulster[62] who is not ready with hundreds and thousands of arguments against anything of the kind'. Sinn Féin should be alive now to 'our unflinching determination' not to go into an all-Ireland parliament.

Strangely, the significance of this speech does not appear to have been grasped by Sinn Féin, either in Dublin or London. As late as 5 December, the delegates enter-

57 Jones, op. cit., p.171. A cabinet meeting was held on 22 November. The negotiations were not discussed. **58** Pakenham, op. cit., pp 241, 376–7. **59** Jones, op. cit., p.174. **60** *Dáil cabinet* (25 November 1921) DE 1/3, NAI. The only records of Dáil cabinet meetings are the notes by the cabinet secretary, Diarmuid O'Hegarty (or in his absence, Colm Ó Murchada), now in DE 1/3 NAI. *DIFP*, i, summarizes these meetings in November-December 1921 (pp 317–46). Pakenham's account of these meetings (pp 245, 255–61), based on Robert Barton, accords very closely with the secretaries' notes. **61** *N.I. Parl. Debates*, i (29 Nov. 1921), pp 294–5. **62** Craig was plainly ignoring the Ulster Nationalists.

tained the illusory hope that Craig could be persuaded to write a letter in support of Irish unity, at least in principle.

Lloyd George instructed his draftsmen, headed by Lionel Curtis,[63] to have the final draft of the Treaty ready by 29 November, but it was not presented to the delegation until the following evening. The oath to the king was still there, but only in his capacity as head of the empire. The provision that the British navy exclusively should defend Irish coasts was to be reviewed after ten years. The prohibition of protective duties between Ireland and England remained. This was the draft Treaty that the Irish delegates brought to Dublin for a special meeting with the Sinn Féin cabinet on Saturday 3 December.[64] The delegates were tired, because of delay following an accident at sea on the previous night. At once a division of opinion emerged. Arthur Griffith strongly urged acceptance of the draft Treaty as the best terms available. He refused to break on the issue of allegiance to the crown, as giving the advantage to the Ulster Unionists. He was supported by Collins and Duggan. Collins had contacted his IRB colleagues that morning and found that they objected to the oath.[65] He told the meeting that they should go to the country on the Treaty, but reject the oath. Duggan agreed with Griffith. Barton and Gavan Duffy took the opposite line – against acceptance of the Treaty. They both argued that England was bluffing, that she would not go to war on the question of allegiance; the current terms were unsatisfactory both on status and Ulster.

According to the record, President de Valera said that he could not accept the oath or the right of Ulster to opt out. But as Lawlor points out, de Valera's attitude was far from straightforward. He was prepared to accept the Treaty with 'modifications' and urged the delegates to go back and continue the negotiations. He proposed an alternative form of oath, giving allegiance to the Constitution of Ireland, to the Treaty of Association, and to the king as head of the Association.[66] Lawlor surmizes that it was 'most likely' that de Valera expected the negotiations to continue indefinitely, but if a break came it should be on the Ulster issue.[67]

The meeting dragged on all day; two plenary sessions separated by a meeting of the Dublin cabinet. Not only the London delegates were tired; de Valera and Brugha had driven from Clare the previous evening.[68] The republican ministers Brugha and Stack displayed intense bitterness towards what they saw as the weakness of Griffith and Collins, who in turn resented this attitude. Brugha made the offensive remark that the British government selected its men. When Griffith protested, Brugha withdrew the charge, but the damage had been done.

To the end Griffith refused to take responsibility for a break with the crown. When as many concessions as possible had been obtained and 'when it was accepted by Craig' he would go before the Dáil, 'the body to decide for or against war'.

63 Lavin, *Curtis*, pp 192–3. **64** For this meeting see *DIFP* i (31 Dec. 1921) pp 344–6. **65** As an oath-bound society the IRB had obvious objections to taking another oath. **66** Pakenham, op. cit., pp 261–2. **67** Lawlor, op. cit., pp 140–1. **68** Longford and O'Neill, *De Valera*, p. 160.

At this point Brugha turned to Griffith and said 'Don't you realise that if you sign this thing, you will split Ireland from top to bottom?', to which Griffith replied, 'I suppose that's so' and promised that he would not sign the document, but bring it back and submit it to the Dáil and, if necessary, to the people.[69] This assurance appears to have satisfied everybody.[70]

The meetings, which had lasted seven hours, ended just in time for the delegates to catch the boats to London. They were told to carry out their original instructions with the same powers; to return to London and say that the cabinet would not accept the oath of allegiance, if not amended, and to face the consequences, 'assuming that England will declare war'. Griffith was personally instructed to tell Lloyd George that the document could not be signed, that it was now a matter for the Dáil, and 'to try to put the blame on Ulster'.[71]

Back at the delegation's London residence (22 Hans Place),[72] the republicans, Barton and Gavan Duffy with Childers' assistance prepared a memorandum of amendments to the proposed Treaty in the light of the cabinet decisions on the previous day.[73] All powers in Ireland, legislative, executive and judicial, should be derived exclusively from 'the elected representatives of the Irish people'. External association reappeared, together with the revised oath proposed by de Valera. Customs duties were to be abolished between Ireland and Great Britain. The naval facilities required by the British were conceded and there was a conditional acceptance of liability for part of the national debt and war pensions, but there was no reference to dominion status or allegiance.

The document was discussed by the entire delegation at 3 o'clock on Sunday 4 December. At once dissension broke out. Griffith, Collins and Duggan refused to present the amendments, since the British would never accept them and they would not take the responsibility for breaking off the negotiations.[74] But when Barton and Gavan Duffy said they would go alone to a meeting arranged for 5 o'clock Griffith agreed to lead the Irish delegation. Neither Collins nor Duggan accompanied him, Collins afterwards writing that he had 'in [his] own estimation argued fully all points'.[75] So just three Irish delegates met Lloyd George, Chamberlain, Birkenhead and Sir Robert Horne (chancellor of the exchequer) at 5 o'clock.

The meeting was a disaster. Griffith, as he afterwards wrote to de Valera, tried to get the discussion focussed on the Ulster issue, 'but could not get it into its proper place'.[76] He stated that the Irish did not take any responsibility for the Ulster pro-

69 Robert Barton and Austin Stack were the primary sources for this exchange, which is not recorded in the cabinet minutes. Pakenham (op. cit., p. 260) was the first secondary work to refer to it. Its accuracy was not challenged by Maye, *Griffith*, p.227.   **70** Pakenham, op. cit., p. 260.   **71** *DIFP* i (3 Dec. 1921), p.246.   **72** All except Collins, who stayed at Cadogan Gardens.   **73** *DIFP* i (4 Dec. 1921), pp 346–8.   **74** Pakenham, pp 264–7.   **75** Memorandum of interview between Collins and Lloyd George (5 Dec. 1921), *DIFP*, i, p.350.   **76** Griffith to de Valera (4 Dec. 1921), *DIFP*, i, pp 348–9. (This was the last letter between the two during the negotiations.)

posals. ('They were theirs, not ours.') He said that he had written them a letter, conditionally accepting association with the empire and ' a recognition of the crown in exchange for essential unity'. Craig should now write a letter accepting essential unity. The British replied that Craig would not write such a letter, but they could nevertheless go ahead with the Treaty.

When Griffith said that the Irish were prepared to accept the Treaty with the proposed amendments, Lloyd George (after an interval) replied that the amendments were a complete repudiation of the discussions of the previous week. They had been offered membership of the empire, a fundamental condition, like the Boer republics which had fought equally gallantly. Griffith vainly pointed out that they were giving up their best ground without even a guarantee from Craig. The British asserted that their own dominions would denounce them, if they even considered the Irish proposals.

When the British asked what was the difficulty about going into the empire, like Canada, Gavan Duffy unguardedly said that the Irish should be as closely associated with Britain as the dominions, 'in the large matters' and more so in the matter of defence, 'but our difficulty is coming within the empire'. At that Chamberlain jumped up and said, 'That ends it.' The discussion ended with the British threatening to inform Craig on the following day that the negotiations had broken down.

Driving back to Hans Place, Griffith 'turned on Gavan Duffy, the proximate occasion of the catastrophe and rent him',[77] while Curtis and Grigg at Lloyd George's behest prepared a statement justifying the breakdown in the negotiations. The statement in Curtis's elegant prose was handed in during the night to the watchman at 10 Downing Street to be ready for the prime minister at 7 a.m. on the following morning, 5 December:[78]

> After five months of negotiation and nearly two months of conference, in the course of which with the utmost patience His Majesty's government have searched for any clue which could lead to a settlement, the Irish representatives have nothing to offer them but proposals which would break the empire in pieces, dislocate society in all its self-governing nations and cancel for ever the hope of national unity in Ireland itself.

Accordingly, the government felt that no useful purpose could be served by continuing the conference. Unless the draft agreement (enclosed for public information) was accepted by (blank date), the government would have no choice but to apply the provisions of the government of Ireland Act. It looked like the end of negotiations, but Tom Jones was determined to save them, if at all possible. At 12.30 a.m. on Monday 5 December he called on Griffith at Hans Place and 'struggled' to persuade him to get Collins to see Lloyd George personally on the following morning. In an unusually emotional letter Jones assured the prime minister, 'One was bound to feel that to

---

77 Pakenham, op. cit., p. 270.   **78** Jones, op. cit., p. 181. This draft is not quoted in Lavin, *Curtis.*

break with him would be infinitely tragic.'[79] Griffith stated that he and Collins had been 'completely won over' to belief in Lloyd George's desire for peace, but they had not persuaded their Dublin colleagues. Unless they could take back something to the Dáil, about half the deputies would vote against them and '90% of the gun-men would follow Collins'. Could not the prime minister, Griffith pleaded, get from Craig a conditional recognition of Irish unity in return for the acceptance of the empire by Sinn Féin? Jones did not comment on whether this request was even remotely feasible, but ended by writing that (presumably at Griffith's request) Collins would call on the prime minister at 9.15 a.m. The letter ended, 'War is failure.'

For Lloyd George the essential issue was the empire; for Collins, the position of the 'North-East', which he found quite unsatisfactory. During his reluctant interview with Lloyd George, Collins made the reasonable request that there should be no repetition of the recent incident in which the Northern government forcibly closed the offices of the democratically-elected Tyrone county council – which might lead to a conflict throughout Ireland. Lloyd George said Collins could raise the matter with Craig, which wasn't much of a reassurance. The interview lasted half an hour; then Lloyd George had a meeting with the king.

At noon the British cabinet met for the first serious discussion of the Irish negotiations. There was a good attendance, including the three delegates (Worthington-Evans, Hewart, and Greenwood) who had been almost entirely excluded from the negotiations – but no complaint was recorded. Lloyd George reported on the British 'offer', which would give Ireland full dominion status subject to one or two modifications, especially with regard to the navy. Although there would be an oath, there would be no veto by Great Britain upon any Irish legislation. Northern Ireland would be given between six and twelve months after the Agreement to opt out, in which case provision was made for a readjustment of the boundary between 'Ulster' and the rest of Ireland. He reported that a majority of the members of the Irish cabinet had rejected the terms and their alternative proposals showed that they had no intention of coming within the British empire, but wished to remain an independent republic associated with the empire for certain purposes and owing no allegiance to the king. The cabinet were informed that 'Mr Arthur Griffiths [sic] and Mr. Michael Collins were greatly disappointed at the rejection of the British proposals, while Collins would have preferred an immediate decision on Ulster.'

While agreeing that the form of the oath might be changed to suit the Irish, if the latter agreed to the constitutional provisions, and especially allegiance to the king, the ministers suggested that at the meeting with the Irish delegates in the afternoon, the British should try to secure a settlement including the essentials – allegiance, membership of the empire, and naval facilities.[80]

The second last sub-conference began at 3 p.m. with just seven delegates present – the British 'Big Four' and Griffith, Collins and Barton. Lloyd George began

---

79 Jones, op. cit., pp 180–1.   **80** CAB 23/27, 5 Dec. 1921.

by saying that he must know once and for all where the Irish stood as regards the Ulster proposals. Griffith and Collins argued that they must have a statement from Craig either accepting or rejecting unity. Collins said that if Craig refused inclusion in an all-Ireland parliament 'we should be left in the position of having surrendered our position without … having even secured the essential unity of Ireland'.[81]

Lloyd George then got excited and claimed that the Irish were trying deliberately to bring about a break on Ulster, because 'our people in Ireland' had refused to come within the empire, and that Griffith was letting him down. He produced the famous letter, claiming Griffith had agreed to its contents.[82] Griffith accepted his undertaking, but argued that it was not unreasonable to require a reply from Craig 'before we refused or accepted the proposals now before us'. Lloyd George replied that this was still letting him down, because the only alternative to Craig's acceptance was the Boundary commission, which his government would implement with strict fidelity. At this, Griffith capitulated. He agreed he would personally sign the Treaty, whether Craig accepted or not, but his colleagues were not party to the promise not to let Lloyd George down. Lloyd George now took advantage of Griffith's weak position. He said that they were all plenipotentiaries; it was a matter of peace or war, and every delegate must sign the document and recommend it or there was no agreement. The government had hazarded their political future 'and we must do likewise and take the same risks'.

'With superb artistry' Lloyd George offered the Free State full fiscal autonomy, but insisted on a reply before ten o'clock that evening, and produced the two letters to Craig, one announcing the success of the negotiations and the other their failure. The Irish delegates secured some further minor concessions; a review of the coastal defence issue within five years and a minor alteration to the oath. The agreement would allow a twelve-month transitional period to Northern Ireland, but the delegates insisted on a reduction to one month. Lloyd George considered that a month did not give the Ulster people sufficient time to reflect, but he was prepared to accept the alteration. The delegates then withdrew and were to reassemble at ten o'clock, but did not meet until 11.15.

From the outset of the final discussion at Hans Place it was clear that Griffith felt obliged to accept the Treaty, while Barton and Gavan Duffy were firmly opposed. Duggan would follow Griffith, as always, so the crucial vote was that of Collins. If he sided with the two republicans, the negotiations would be at an end, but he would carry the responsibility in Ireland, and 'he was resolved not to be a scapegoat'. He also mentioned in the discussion that there had been very few active Volunteers, and he plainly did not wish a resumption of hostilities.[83] Historians generally agree that Collins' decision was the decisive one. A dissenter is Iain McLean in his recent *Ratio-*

---

81 The records of the two last sub-conferences are found in the notes by Robert Barton – *DIFP* I, 351–6 (NAI, DE 2/304/1) **82** The text of the letter is in Appendix 2. See also above, p. 125. **83** J.J. Lee, 'The Challenge of a Collins Biography' in G. Doherty & D. Keogh, *Michael Collins and the Making of the Irish State* (Cork, 1998), p. 32.

nal Choice in British Politics (pp 173–203), who suggests that Barton's was the decisive vote ('the decision of his life'). But was it plausible that a Protestant landowner from Wicklow and former British Army officer would stand out against the decision of two representatives of Catholic Ireland?[84] Once Barton signed, Gavan Duffy, the last to sign, had no alternative. All historians have remarked on the extraordinary failure of the delegates to telephone the cabinet in Dublin, as their instructions would have obliged them to. But the silence is not referred to in Childers' diary, and Barton wrote in 1975, shortly before his death, that 'the telephone may have been eschewed because we suspected that it was under supervision'.[85]

The final sub-conference lasted from 11.15 p.m. to 2 a.m. The delegates had agreed to accept the Treaty, but tried to get further emendations. They attacked the clause about the governor-general but failed to get it altered. The only other significant alteration was on the clause requiring the summoning of the Southern parliament to act as a provisional administration until the Constitution of the Irish Free State was enacted. The clever lawyer, Birkenhead, immediately drafted a memorandum making it plain that the functions should be transferred to the Provisional government of Southern Ireland. Finally, Lloyd George asked whether the delegation would accept the Articles of Agreement and stand by them in the Dáil. Griffith replied 'We do.' The British and Irish delegations shook hands (for the first time) and the conference ended at 2.20 a.m. on 6 December.

With the predictable exception of the *Morning Post*, the British daily newspapers united in a 'finely orchestrated chorus' of praise for the Treaty on the morning of 7 December.[86] The three Dublin newspapers were similarly ecstatic. The *Irish Independent* declared, 'After the years of agony we have endured, after the many fluctuations of the conference itself, with its moments of tension and anxiety, Ireland will today joyfully receive the news that the long and arduous deliberations have culminated in a Treaty which ensures her the best possible terms that in existing conditions could be secured.' The reaction of the republicans in Dublin was at first one of incredulity that the delegates had disregarded their instructions, followed by anger. On hearing the terms of the Treaty on 6 December from the *Evening Mail*, de Valera spent 'one of the most miserable evenings of his life, pondering over the news which he had received and trying to keep himself from showing any emotion'.[87] The reaction of Brugha and Stack was similar. On the morning of 7 December de Valera proposed that the four ministers still in Dublin dismiss the three ministers who had signed the Treaty – Griffith, Collins, and Barton – but W.T. Cosgrave for the first time played a decisive role, requesting that a decision be postponed until the delegates returned from London. Many years later, Cosgrave told his son Liam that it

---

84 MacLean may be relying too heavily on the Childers papers. He does not mention Griffith's persistent (and unjust) though something understandable antipathy to Childers. **85** Barton MS TCD 7834/10, quoted in McLean, p. 178. **86** See list of extracts from London dailies in Pakenham, op. cit., p. 326. **87** Longford and O'Neill, *De Valera*, p. 167.

was one of the most significant cabinet meetings he had ever attended.[88] The crucial meeting with all seven members present took place on the following day. Four members – Griffith, Collins, Barton and Cosgrave – voted to recommend that the Dáil should accept the Treaty. De Valera, Brugha, and Stack were opposed. On the 9th the president issued a proclamation informing the public that he could not recommend acceptance of the Treaty to the Dáil or the country, but by that time Irish public opinion had moved towards acceptance. The extent of the division in the Sinn Féin movement was to become clear in the Dáil debates on the Treaty, starting on 14 December.

Like the public representatives, historians have been divided over the Treaty.[89] The leading representatives of both viewpoints are Frank Pakenham and Nicholas Mansergh.[90] Pakenham does not question the *bona fides* of the delegates, but states that they 'had met wills as strong, intelligences as subtle, and diplomacy more experienced than theirs'.[91] Mansergh is more subtle, inferring that once Northern Ireland had been established it was impossible to unscramble it. 'Though much had been gained, neither associate republican status nor unity had been won. In the shorter run the door was left ajar for the second, but not for the first: in the longer run, for the first, but not the second.'[92]

Generally speaking, historians have been more understanding of the Irish difficulties since the publication of Tom Jones's *Whitehall Diary* in 1971. It is clear that the Sinn Féin delegates in London, as well as their colleagues in Dublin, failed to achieve their two main objectives; the sovereignty and unity of Ireland. We do not entirely agree with the quotation from Professor Murphy at the head of this chapter. Griffith and Collins were prepared to trade sovereignty for unity, but unquestionably failed to secure either. The later rationalization of the Treaty ('freedom to obtain freedom') was a second-best. However, there was one victor in the Treaty negotiations – Sir James Craig. He and his party secured their two objectives – the maintenance of the partition settlement secured by the Government of Ireland Act, and exclusion from any meaningful all-Ireland institutions. It seems that at the time Craig did not realise the full extent of his success,[93] but that success was pregnant with trouble for the future.

88 S. Collins, *The Cosgrave Legacy* (Dublin, 1996) p. 24. **89** B. Maye, *Arthur Griffith,* pp 220–42 gives an exhaustive list of the historians on both sides. **90** Interestingly, both originated from the Protestant landed gentry. (Pakenham became a Catholic in 1940). **91** Pakenham op. cit. p.311. **92** Mansergh, *The Unresolved Question* p.189. **93** He wrote to Austen Chamberlain threatening that if the Boundary commission sacrificed any of the territory of 'Ulster' they might be resisted by force. Chamberlain replied, 'I cannot believe that men whose loyalty is their pride are contemplating acts of war against the King': Craig to Austen Chamberlain 15 December 1921, Austen Chamberlain to Craig 16 December 1921, covering letter from Chamberlain to Bonar Law 16 December 1921, all in BLP 107/1/98. (Chamberlain sent copies of the correspondence to Bonar Law in the hope that he could restrain Craig.)

Correspondence

between his majesty's government and the prime minister of Northern Ireland relating to the proposals for an Irish settlement

1. Letter from the Prime Minister to Sir James Craig

10, Downing Street
London, S.W.1,

10th November 1921.

My dear Prime Minister,

(1) The time has arrived when formal consultation between His Majesty's Government and the Government of Northern Ireland is necessary for the further progress of the Irish negotiations. The settlement towards which His Majesty's Ministers have been working, and which they believe is now attainable, is closely based upon the proposals made by His Majesty's Government on the 20th July last.

(2) Such a settlement would comprise the following main principles:-

(a) Ireland would give her allegiance to the Throne and would take her place in the partnership of Free States comprised in the British Empire.

(b) Provision would be made for those naval securities which His Majesty's Government deem indispensable for Great Britain and her overseas communications.

(c) The Government of Northern Ireland would retain all the powers conferred upon her by the Government of Ireland Act.

(d) The Unity of Ireland would be recognised by the establishment of an all-Ireland Parliament, upon which would be devolved the further powers necessary to form the self-governing Irish State.

(3) Northern Ireland will no doubt see many dangers in a settlement on these lines. It may be objected, for instance, that the patronage involved in the various common Departments, such as the Post Office, Customs and Excise, might be unfairly exercised on religious and other grounds; or again, that though Ulster would retain control of its education and kindred subjects, the Irish Government would be in a position to withhold the funds necessary to defray the administrative cost. Moreover, it might be feared that if the all-Ireland Parliament were to control import and export trade, the industries of Ulster would be imperilled.

(4) His Majesty's Government recognise the force of these objections, and desire to consider, in consultation with yourself and your Cabinet, how they can best be met. They invite your Cabinet to discuss with them the best means of dealing with these and similar matters; in particular (a) the appointment of officials within the area of Northern Ireland in Departments subject to the all-Ireland Parliament; (b) the collection of revenue within the area of Northern Ireland; (c) measures for safeguarding the exports and imports of Northern Ireland against the imposition of discriminating duties and its citizens from invidious taxation.

His Majesty's Ministers believe that arrangements can be embodied in the Agreement now in view, whereby these difficulties can be met.

(5) The question of the area within the special jurisdiction of the Northern Parliament we have reserved for discussion with you. The creation of an all-Ireland Parliament would clearly further an amicable settlement of this problem.

(6) His Majesty's Government are fully aware of the objections which the people of Northern Ireland may feel to participation on any terms in an all-Ireland Parliament. They have, therefore, been examining some of the alternatives, and their consequences. Their study has convinced them that grave difficulties would be raised for both parts of Ireland, if the jurisdiction over the reserved subjects were not conferred upon a common authority.

In the first place Customs barriers would have to be established between Northern and Southern Ireland over a jagged line of frontier. The inconvenience of this arrangement would be considerably enhanced by the fact that there must of necessity be large elements of the population on both sides whose sympathies will lie across the border. The natural channels of trade would be arbitrarily obstructed. The difficulty of working any such arrangement would be unceasing, the cost considerable, and the vexation to traders continuous.

In the second place the finance of the Government of Ireland Act would necessarily have to be re-cast. It is the essence of Dominion status that the contribution of a Dominion towards Imperial charges is voluntary. If Northern Ireland were part of the Irish State its contribution would be voluntary, like those of the Dominions. On the other hand, if Southern Ireland became a Dominion while Northern Ireland remained a part of the United Kingdom with the essential corollary of representation in the Imperial Parliament, it is clear that the people of Northern Ireland would have to bear their proportionate share of all Imperial burdens, such as the Army, Navy and other Imperial Services, in common with the taxpayers of the United Kingdom. The Members for Northern Ireland at Westminster would otherwise be voting for policies in Parliament, the expense of which they would not share. It would be inevitable, if Northern Ireland were to remain a part of the United Kingdom, for Belfast to bear the same burdens as Liverpool, Glasgow or London.

These illustrations are by no means exhaustive, but they are sufficient to show the kind of difficulties which must arise from the grant of Dominion powers to a part of Ireland only.

(7) It will be evident, that the people of Great Britain are making important sacrifices for the sake of a settlement. Heavily burdened though Great Britain is, the Government,

with the full consent of public opinion at home and throughout the Empire, is offering to forego her right to exact from Ireland any contribution to future Imperial expenses. Single handed, the British Nation assumes responsibility for Imperial Defence, except in so far as Ireland and the Dominions may resolve of their own free-will to contribute to the cost.

(8) His Majesty's Government have purposely reviewed the problem in broad outline only. The details of any settlement cannot be satisfactorily approached except by discussion between all parties concerned. It is not possible by correspondence to deal adequately with even the main features of the question as it now stands, and His Majesty's Government cordially invite the Ministers of Northern Ireland to meet them in conference with a view to a full and frank exchange of views.

Yours sincerely,
(Signed) D. Lloyd George.
The Rt. Hon. Sir James Craig, Bart.

---

2. Letter from Sir James Craig to the Prime Minister

Constitutional Club,
Northumberland Avenue,
W.C.

6 p.m., November 11th, 1921.

My dear Prime Minister,

(1) The outline of proposals towards a settlement of the Irish question submitted by you on behalf of His Majesty's Government in your communication of the 10th instant has been carefully considered by me in consultation with my colleagues in the Government of Northern Ireland, on whose behalf I have the honour to submit to you the following considerations in reply thereto.

(2) (a) The question of giving their allegiance to the Throne does not arise in the case of the people of Ulster, as they have always been amongst His Majesty's most loyal and devoted servants, but they will gladly embrace any opportunity that may be afforded them of emphasising afresh their loyalty to His Majesty's Crown and Person, which was so signally displayed on the occasion of the opening of the Parliament of Northern Ireland by His Majesty in June last. It has always been the desire of Northern Ireland to remain in the closest possible union with Great Britain and the Empire, which Ulstermen have helped to build up, and to which they are proud to belong.

The Government of Northern Ireland feel constrained to observe that it is with surprise they find the question of allegiance to the Throne, and membership of the British Empire included among the heads of your proposals, inasmuch as it has more than once been unequivocably stated by yourself in published correspondence with the representatives of Southern Ireland, that these two fundamentals were not open for discussion. The Government of Northern Ireland, having made their own position clear, hold that the attitude of Southern Ireland towards these points is a matter which, as it affects the solidarity of the Empire, rests with His Majesty's Ministers to decide. At the same time it would, of course, be a matter of satisfaction to the Government of Northern Ireland and to the loyal population they represent to feel assured that the permanent allegiance of the people of Southern Ireland to His Majesty and their enduring participation in the partnership of the British Empire were no longer in question.

(b) Not only does Northern Ireland assent to provision being made for those Naval Services which His Majesty's Government deem indispensable for Great Britain and her overseas communications, but will be ready at any time to co-operate to the utmost of her ability in any measures that may be taken for such purpose.

(c) As a final settlement and supreme sacrifice in the interests of peace, the Government of Ireland Act, 1920, was accepted by Northern Ireland, although not asked for by her representatives. My colleagues and I are surprised that you should think it necessary to emphasise the fact that you do not propose to take away any of the powers given to us so lately as last year. We observe, with considerable concern, notwithstanding this assurance that in paragraph 5 of your communication, the area within the jurisdiction of the Northern Parliament is referred to as being open to possible revision. This question was very carefully and fully considered in all its aspects, when the Government of Ireland Act was under discussion in Parliament last year. The area finally decided upon is defined in the Act, and it forms no less essential a part of the Act than the powers conferred upon the Northern Parliament.

(d) The possible unity of Ireland is provided for by the establishment of the Council of Ireland under the Act of 1920, together with the machinery for creating a Parliament for all-Ireland should Northern and Southern Ireland mutually agree to do so. The proposal now made to establish an all-Ireland Parliament by other means, presupposes that such agreement is not necessary. An all-Ireland Parliament cannot under existing circumstances be accepted by Northern Ireland.

Such a Parliament is precisely what Ulster has for many years resisted by all the means at her disposal, and her detestation of it is in no degree diminished by the local institutions conferred upon her, by the Act of 1920. The Government of Northern Ireland deem it unnecessary to enumerate here the reasons for this repugnance of which as stated in paragraph 6, of your communication His Majesty's Ministers are fully aware; but they must observe that the objection of Northern Ireland to participation in an all-Ireland Parliament, so far from being weakened, has been materially strengthened by recent events in other parts of Ireland to which it is unnecessary more particularly to refer.

It is an objection that goes deeper then the consideration referred to in paragraph 3 of your communication.

The Government of Northern Ireland consider it their duty in order to avoid misunderstanding to say that their inability to accept an all-Ireland Parliament does not depend merely on the question of safeguards in regard to administrative details such as those referred to in paragraphs 3 and 4 of your communication. They are certain that no paper safeguards could protect them against maladministration. The feelings of the loyal population of Ulster are so pronounced and so universal on this point that no Government representing that population could enter into any conference where this point is open to discussion. For these reasons, therefore, they feel compelled to state that any discussion would be fruitless unless His Majesty's Ministers consent to the withdrawal of the proposal for an all-Ireland Parliament.

(3) The Government of Northern Ireland are fully alive to the difficulties referred to in paragraph 6 of your communication, but they cannot agree that the only way, or the best way, of surmounting those difficulties under existing circumstances is by conferring jurisdiction over the reserved subjects upon a common authority. His Majesty's Ministers assume that the only alternative to such an arrangement is that, while the status of a Dominion should be given to Southern Ireland, Northern Ireland would remain a part of the United Kingdom with the essential corollary of representation in the Imperial Parliament, and certain financial and other disadvantages which Northern Ireland, as compared with Southern Ireland, would suffer under such an arrangement are pointed out.

There is another alternative which His Majesty's Ministers do not appear to have considered. It is that the reserved powers instead of being entrusted to a common authority should be conferred on the Government of Southern and of Northern Ireland within the areas of their respective jurisdictions. The principle underlying the Government of Ireland Act, 1920, was equality of status and of powers for the two Governments in Ireland, and this principle should, in the opinion of the Government of Northern Ireland, be observed in the transfer of reserved services.

If the plan here suggested were followed, it would obviate the chief difficulty referred to in paragraph 6 of your communication.

It is true that it might involve Northern Ireland losing her representation in the Imperial Parliament; but while Northern Ireland would deplore any loosening of the tie between Great Britain and herself she would regard the loss of representation at Westminster as a less evil than inclusion in an all-Ireland Parliament. It is realised that if the alternative here suggested were adopted, the contribution of Northern Ireland, as also of Southern Ireland, to the cost of Imperial Services would be voluntary, as in the case of the Overseas Dominions. The proved loyalty of Northern Ireland to the British Empire is a sufficient guarantee that she would not evade this obligation, and the Government of Northern Ireland are convinced that by this plan the Imperial Exchequer would have better security for a contribution from Ireland to the cost of Imperial Services than if it depended on a voluntary contribution from an all-Ireland Parliament, the majority of whose members would not be likely to be animated by sentiments of affection for Great Britain.

It will be seen, therefore, that the Government of Northern Ireland are prepared to accept three out of the four proposals put forward by His Majesty's Government. The fourth proposal they are unable to accept for the reasons stated, and they respectfully invite the attention of His Majesty's Government to the alternative suggestions here submitted.

In conclusion, the Government of Northern Ireland desire to express their firm conviction that the time has not yet arrived when the cause of peace in Ireland, which they fervently desire to further by all means in their power, can be promoted by establishing an all-Ireland Parliament. Such a Constitution can only come from mutual confidence, and when the time for it comes the provisions of the Act of 1920 will prove sufficient for the purposes of bringing it into existence.

Yours sincerely,
(Signed) James Craig.
The Rt. Hon. David Lloyd George, O.M., M.P.

---

3. Letter from the Prime Minister to Sir James Craig

My dear Prime Minister,

14th November, 1921.

We have received with great regret your refusal to enter into conference with us unconditionally. To demand as between two sets of Ministers of the Crown a preliminary limitation on freedom of discussion is contrary to the spirit of mutual loyalty and co-operation which animates His Majesty's Governments in all parts of the Empire.

We regret it the more because your counter-proposal that Southern and Northern Ireland should be constituted two separate Dominions is in our judgment indefensible.

We are opposed to it, in the first place, on the ground of broad Imperial principle. To create two Dominions in Ireland, one of twenty-six and one of six counties, would fundamentally change the existing system of Imperial Organisation. The status of the Dominions, both nationally and internationally, is based upon the gradual amalgamation of large territories and scattered colonies in natural units of self-government. We could not reasonably claim place for two Irelands in the Assembly of the League of Nations or in the Imperial Conference. If Ireland is represented in either institution, it must be preferably Ireland as a whole or, failing the whole, that part of it which has the largest population and area. To demand the same national and international status for six counties separately is a proposal which we could not reconcile with the Empire's internal and foreign interests.

Your proposal would, in our opinion, be equally injurious from the domestic standpoint of the British Isles, both financially and commercially. The erection of two systems of national government in these islands is sufficiently beset with difficulties. His Majesty's Government have determined to face these difficulties for the sake of peace at the heart of the Empire, and the ultimate unity of Ireland. Neither of these objects would be served by the erection of three national governments, involving three systems of customs and excise, three rates of income tax, and three currencies. The injury of such a treble system to the trade of Great Britain would be considerable; to that of Ireland it would be ruinous. That the business community of Northern Ireland would endorse such a proposal when once they had realised its implications, appears to us inconceivable.

All experience proves, moreover, that so complete a partition of Ireland as you propose must militate with increasing force against that ultimate unity which you yourself hope will one day be possible. The existing state of Central and South-Eastern Europe is a terrible example of the evils which spring from the creation of new frontiers, cutting the natural circuits of commercial activity; but when once such frontiers are established, they harden into permanence. Your proposal would stereotype a frontier based neither upon natural features nor broad geographical considerations by giving it the character of an international boundary. Partition on these lines the majority of the Irish people will never accept, nor could we conscientiously attempt to enforce it. It would be fatal to that purpose of a lasting settlement on which these negotiations from the very outset have been steadily directed.

We cannot, finally, overlook the effect of your proposal upon the welfare of the minorities both in Southern and Northern Ireland. In both parts of Ireland these are considerable communities cut off from the majority of those to whom they are bound by faith, tradition and natural affinity. The majority in Southern Ireland have a strong sense of responsibility for their co-religionists in the six counties. The minority there have an equal interest in your sympathy and support.

The considerations which I have outlined make free inter-change of ideas between us essential, and we sincerely trust that you will not persist in making preliminary conditions upon matters which can only be properly explored in conference. I hope, therefore, that you will come and see me at your earliest convenience.

I am,
Yours sincerely
(Signed) D. Lloyd George.
The Right Hon. Sir James Craig, Bart.

---

4. Letter from Sir J. Craig to the Prime Minister

Constitutional Club,
Northumberland Avenue.

17th November 1921.

My dear Prime Minister

In your letter of the 14th inst. you express regret that my colleagues and I have found it impossible to meet you in formal Conference so long as your proposal that we should agree to the establishment of an "ALL Ireland Parliament" was open to discussion.

To enter the Conference on such a condition would in our view be dishonest, since we know that in no circumstances would Ulster accept such a position, involving permanent subordination to Sinn Fein. We are strongly convinced that it could only tend to make settlement more difficult and encourage false hopes, if even by implication we discussed a condition which cannot be conceded.

In your letter of the 10th inst. you indicated an alternative but made no mention of the course suggested in our reply. We urged that if you are resolved upon the complete fiscal separation of Ireland from the rest of the United Kingdom, the same equality of treatment as between Northern and Southern Ireland should be maintained in dealing with the reserved services as has been pursued with regard to the Services already transferred under the Act of 1920.

You now mention various objections to our proposal and suggest that Ulster would be led to economic ruin if she were separated from Southern Ireland. You have apparently over-looked the fact that your proposal to break the fiscal unity of the United Kingdom would involve the fiscal separation of Ulster from Great Britain, with which 90 per cent. of her trade is – directly or indirectly – conducted. Can it be doubted that more harmonious trade relations will result between Great Britain and Ulster by the control of these matters being in our own hands rather than in the hands of an All-Ireland Parliament dominated by Sinn Fein, which during the past twelve months has enforced a trade boycott against Northern Ireland? As to the question of currency, we view with grave concern your proposal to establish a separate Irish currency. We are confident that Commercial and banking interests will refuse to endorse such a proposal once they realise its implication.

We concur with the view that the creation of new frontiers tends to harden into permanence. Why then seek to establish such a frontier between Great Britain and Ireland? If, however, you are determined upon such a policy, is it not better to grant to Ulster the status of a separate Dominion and thus insure a firm and abiding link between Northern Ireland and the Mother Country? That being granted, we would be most happy to entrust to Great Britain the safeguarding of our common interests in the Imperial Conference and the League of Nations.

You point out the great difficulties of creating a Northern Dominion in Ireland, and you refer to this as our counter proposals. That description is not quite accurate. We feel that the arguments you use as objections to two Dominions apply with equal force to the creation of one. We only put forward our suggestions because we are convinced that if you once violate the fiscal unity of the United Kingdom it makes relatively little difference to create two new units instead of one.

Your proposal involved the placing of Ulster under Sinn Fein, which is an insurmountable difficulty. I desired to be helpful by pointing out a method by which if this concession is to be made to the South and West it can also be made to the North without creating the grave results I have indicated. But you must not argue from that that we in the slightest degree modify our convictions that your own proposals embodied in the Act of last year are the only safe and sound plan both for Great Britain and for Ireland.

To sum up:- If you force Ulster to leave the United Kingdom against the wishes of her people, she desires to be left in a position to make her own fiscal and international policy conform as nearly as possible with the policy of the Mother Country, and to retain British traditions, British currency, British ideals, and the British language, and in this way render the disadvantages entailed by her separation from Great Britain as slight as possible.

We are resolved to set such an example of good government and just administration within our jurisdiction, as shall inspire the minority in our midst with confidence, and we hope lead eventually to similar conditions being established throughout the rest of Ireland.

Our position having been made perfectly clear in this and our former letter, if you hold the opinion that any good purpose can be served by my seeing you for the interchange of ideas, I shall be at your disposal when I return from the dedication of the Ulster Battlefield Memorial at Thiepval on Monday next.

In conclusion my colleagues and I desire again to represent with all respect that in our opinion it is of great importance that full publicity shall be given to our respective views forthwith, so as to put an end at once to the campaign of misrepresentation in the Press to which Mr. Austin Chamberlain equally with ourselves takes great exception.

Yours sincerely

(Signed) James Craig

The Right Hon. D. Lloyd George O.M., M.P.

---

5. Letter from the Prime Minister to Sir James Craig

10, Downing Street,
S.W.1

18th November 1921.

Dear Mr Craig,

I am sorry to see from the papers that you are suffering from an attack of influenza. I can only hope that it is not a severe attack and that we shall soon hear of your complete recovery. In these circumstances I do not propose to trouble you with a lengthy answer to your letter of yesterday. When you are fully recovered let us meet, as you suggest, at the end of that letter for an informal talk. We can then see how to get over the difficulties which seem to stand in the way of a Conference free from all conditions.

I should be greatly obliged if you would leave over the question of publication until we meet. There are obvious difficulties which ought to be discussed fully between us before any decision is taken.

I am
Yours sincerely
(Signed) D. Lloyd George
The Right Hon. Sir James Craig, Bart.

6. Letter from Sir James Craig to the Prime Minister

Grosvenor Hotel
S.W. 1

20the November, 1921.

My dear Prime Minister

Very many thanks for your letter of the 18th instant, and for your kind enquiries after my health. I am better and hope to be able to return to Ulster on Thursday night.

I shall be glad to meet you for an informal conversation before I go back, any time you may appoint on Wednesday.

I readily agree to defer publication until after our meeting, but in view of the Assembly of the Parliament of Northern Ireland of the 29th instant, you will, I feel sure, agree with me that publication is essential before that date, and that all the correspondence should be published, beginning with your first invitation to me and my reply thereto. By this it will be made quite plain to the public – that it is not on our part that there has been the refusal to enter into Conference with you, but that it is the Sinn Fein delegates who have refused to let us take part unless we do so in a subordinate position to themselves.

I should also have to let our Parliament be informed whether Sinn Fein was prepared to give allegiance to the Crown without reservation, which was one of the conditions of your invitation, or whether their consent to do so is still withheld and made dependent on your first having procured the consent of Ulster to an All-Ireland Parliament.

Yours sincerely
(Signed) James Craig
The Rt. Hon. D. Lloyd George, O.M., M.P.

---

7. Letter from the Prime Minister to Sir James Craig accompanying the articles of agreement

December 5th, 1921.

My dear Prime Minister

I enclose Articles of Agreement for an Irish Settlement which have been signed on behalf of H.M. Government and of the Irish Delegation. You will observe that there are two alternatives between which the Government of Northern Ireland is invited to choose. Under the first, retaining all her existing powers, she will enter the Irish Free State with such additional guarantees as may be arranged in conference. Under the second alternative she will still retain her present powers, but in respect of all matters not already delegated to her will share the rights and obligations of Great Britain. In the latter case, however, we should feel unable to defend the existing boundary, which must be subject to revision on one side and the other by a Boundary Commission under the terms of the Instrument.

I have only to add that I shall be glad to arrange for an early meeting of the Conference contemplated in Article 15, or for any preliminary or less formal discussion, which you may desire with my colleagues and myself.

Ever sincerely
(Sgd.) D. Lloyd George
the Right Hon. Sir James Craig, Bart.

APPENDIX II

Memorandum[1]

from Griffith to Lloyd George, 12 November 1921

If Ulster did not see her way to accept immediately the principle of a Parliament of All-Ireland – coupled with the retention by the Parliament of Northern Ireland of the powers conferred upon it by the Act of 1920 and such other safeguards as have already been suggested in my letter of 10th November – we should then propose to create such Parliament for All-Ireland but to allow Ulster the right within a specified time on an address to the Throne carried in both houses of the Ulster Parliament to elect to remain subject to the Imperial Parliament for all the reserved services. In this case she would continue to exercise through her own Parliament all her present rights; she would continue subject to British taxation except in so far as already modified by the Act of 1920. In this case, however, it would be necessary to revise the boundary of Northern Ireland. This might be done by a Boundary Commission which would be directed to adjust the line both by inclusion and exclusion so as to make the Boundary conform as closely as possible to the wishes of the population.

---

1 Ronan Fanning, Michael Kennedy, Dermot Keogh and Eunan O'Halpin (eds), *Documents in Irish Foreign Policy*, i, 1919–22 (Dublin, 1998), pp 352–3.

Articles of Agreement

for a treaty between Great Britain and Northern Ireland

1. Ireland shall have the same constitutional status in the Community of Nations known as the British Empire as the Dominion of Canada, the Commonwealth of Australia, the Dominion of New Zealand, and the Union of South Africa with a Parliament having powers to make laws for the peace, order and good government of Ireland and an Executive responsible to that Parliament, and shall be styled and known as the Irish Free State.
2. Subject to the provisions hereinafter set out the position of the Irish Free State in relation to the Imperial Parliament and Government and otherwise shall be that of the Dominion of Canada, and the law, practice and constitutional usage governing the relationship of the Crown or the representative of the Crown and of the Imperial Parliament to the Dominion of Canada shall govern their relationship to the Irish Free State.
3. The representative of the Crown in Ireland shall be appointed in like manner observed in the making of such appointments.
4. The oath to be taken by Members of the Parliament of the Irish Free State shall be in the following form:–

    I.......do solemnly swear true faith and allegiance to the Constitution of the Irish Free State as by law established and that I will be faithful to H.M. King George V., his heirs and successors by law, in virtue of the common citizenship of Ireland with Great Britain and her adherence to and membership of the group of nations forming the British Commonwealth of nations.

5. The Irish Free State shall assume liability for the service of the Public Debt of the United Kingdom as existing at that date hereof and towards the payment of War Pensions as existing at that date in such proportion as may be fair and equitable, having regard to any just claim on the part of Ireland by way of set-off or counter-claim, the amount of such sums being determined in default of agreement by the arbitration of one or more independent persons being citizens of the British Empire.
6. Until an arrangement has been made between the British and Irish Governments whereby the Irish Free State undertakes her own coastal defence, the defence by sea of Great Britain and Ireland shall be undertaken by His Majesty's Imperial Forces, but this shall not prevent the construction or maintenance by the Government of the Irish Free State of such vessels as are necessary for the protection of

the Revenue or the Fisheries. The foregoing provisions of this article shall be reviewed at a conference of Representatives of the British and Irish governments, to be held at the expiration of five years from the date hereof with a view to the undertaking by Ireland of a share in her own coastal defence.

7.   The Government of the Irish Free State shall afford to His Majesty's Imperial Forces

   (a)   In time of peace such harbour and other facilities as are indicated in the Annex hereto, or such other facilities as may from time to time be agreed between the British Government and the Government of the Irish Free State; and

   (b)   In time of war or of strained relations with a Foreign Power such harbour and other facilities as the British Government and the Government may require for the purposes of such defence as aforesaid.

8.   With a view to securing the observance of the principle of international limitation of armaments, if the Government of the Irish Free State establishes and maintains a military defence force, the establishments thereof shall not exceed in size such proportion of the military establishments maintained in Great Britain as that which the population of Ireland bears to the population of Great Britain.

9.   The ports of Great Britain and the Irish Free State shall be freely open to the ships of the other country on payment of the customary port and other dues.

10.   The Government of the Irish Free State agrees to pay fair compensation on terms not less favourable than those accorded by the Act of 1920 to judges, officials, members of Police Forces and other Public Servants who are discharged by it or who retire in consequence of the change of government effected in pursuance hereof. Provided that this agreement shall not apply to members of the Auxiliary Police Force or to persons recruited in Great Britain for the Royal Irish Constabulary during the two years next preceding the date hereof. The British Government will assume responsibility for such compensation or pensions as may be payable to any of these excepted persons.

11.   Until the expiration of one month from the passing of the Act of Parliament for the ratification of this instrument, the powers of the Parliament and the Government of the Irish Free State shall not be exercisable as respects Northern Ireland, and the provisions of the Government of Ireland Act 1920, shall, so far as they relate to Northern Ireland, remain full of force and effect, and no election shall be held for the return of members to serve in the Parliament of the Irish Free State for constituencies in Northern Ireland, unless a resolution is passed by both Houses of Parliament of Northern Ireland in favour of the holding of such elections before the end of the said month.

12.   If before the expiration of the said month, an address is presented to His Majesty by both Houses of the Parliament of Northern Ireland to that effect, the powers of the Parliament and the Government of the Irish Free State shall no longer extend to Northern Ireland, and the provisions of the Government of Ireland Act 1920, (including those relating to the Council of Ireland) shall so far as they relate to Northern Ireland, continue to be of full force and effect, and this instrument shall

have effect subject to the necessary modifications.

Provided that if such an address is so presented a Commission consisting of three persons, one to be appointed by the Government of the Irish Free State, one to be appointed by the Government of Northern Ireland, and one who shall be Chairman to be appointed by the British Government shall determine in accordance with the wishes of the inhabitants, so far as may be compatible with economic and geographic conditions, the boundaries between Northern Ireland and the rest of Ireland, and for the purposes of the Government of Ireland Act, 1920, and of this instrument, the boundary of Northern Ireland shall be such as may be determined by such Commission.

13.   For the purpose of the last foregoing article, the powers of the Parliament of Southern Ireland under the Government of Ireland Act, 1920, to elect members of the Council of Ireland shall after the parliament of the Irish Free State is constituted be exercised by that Parliament.

14.   After the expiration of the said month, if no such address as is mentioned in Article 12 hereof is presented, the Parliament and Government of Northern Ireland shall continue to exercise as respects Northern Ireland the powers conferred on them by the Government of Ireland Act, 1920, but the Parliament and Government of the Irish Free State shall in Northern Ireland have in relation to matters in respect of which the Parliament of Northern Ireland has not power to make laws under the Act (including matters which under the said Act are within the jurisdiction of the Council of Ireland) the same powers as in the rest of Ireland, subject to such other provisions as may be agreed in manner hereinafter appearing.

15.   At any time after the date hereof the Government of Northern Ireland and the provisional Government of Southern Ireland hereinafter constituted may meet for the purpose of discussing the provisions subject to which the last foregoing Article is to operate in the event of no such address as is therein mentioned being presented and those provisions may include;

(a)   Safeguards with regard to patronage in Northern Ireland.

(b)   Safeguards with regard to the collection of revenue in Northern Ireland.

(c)   Safeguards with regard to import and export duties affecting the trade or industry of Northern Ireland.

(d)   Safeguards for minorities in Northern Ireland.

(e)   The settlement of the financial relations between Northern Ireland and the Irish Free State.

(f)   The establishment and powers of a local militia in Northern Ireland and the relation of the Defence Forces of the Irish Free State and of Northern Ireland respectively, and if at any such meeting provisions are agreed to, the same shall have effect as if they were included amongst the provisions subject to which the powers of the Parliament and the Government of the Irish Free State are to be exercisable in Northern Ireland under Article 14 hereof.

16.   Neither the Parliament of the Irish Free State nor the Parliament of Northern Ireland shall make any law so as either directly or indirectly to endow any religion or prohibit or restrict the free exercise thereof or give any preference or impose any disability on account of religious belief or religious status or affect prejudicially the right of any child to attend a school receiving public money without attending the religious instruction at the school or make any discrimination as respects State aid between schools under the management of different religious denominations or divert from any religious denomination or any educational institution any of its property except for public utility purposes and on payment of compensation.

17.   By way of provisional arrangement for the administration of Southern Ireland during the interval which must elapse between the date hereof and the constitution of a Parliament and Government of the Irish Free State in accordance therewith, steps shall be taken forthwith for summoning a meeting of members of Parliament elected for constituencies in Southern Ireland since the passing of the Government of Ireland Act, 1920, and for constituting a provisional Government, and the British Government shall take the steps necessary to transfer to such provisional Government the powers and machinery requisite for the discharge of its duties, provided that every member of such provisional Government shall have signified in writing his or her acceptance of this instrument. But this arrangement shall not continue in force beyond the expiration of twelve months from the date hereof.

18.   This instrument shall be submitted forthwith by His Majesty's Government for the approval of Parliament and by the Irish signatories to a meeting summoned for the purpose of the members elected to sit in the House of Commons of Southern Ireland and if approved shall be ratified by the necessary legislation.

(Signed)

On behalf of the British Delegation,   On behalf of the Irish Delegation.

D. LLOYD GEORGE   ART Ó GRÍOBHTHA (ARTHUR GRIFFITH)
BIRKENHEAD   MICHAEL Ó COILÉAIN
WINSTON S. CHURCHILL   RIOBÁRD BARTÚN
L. WORTHINGTON-EVANS   E.S. Ó DUGAIN
HAMAR GREENWOOD   SEÓRSA GHABHÁIN UÍ DHUBHTHAIGH
GORDON HEWART

6th December 1921.

ANNEX.

The following are the specific facilities required:–

Dockyard Port at Berehaven.

(a)   Admiralty property and rights to be retained as at the date hereof. Harbour defences to remain in charge of British care and maintenance parties.

Queenstown

(b)   Harbour defences to remain in charge of British care and maintenance parties. Certain mooring buoys to be retained for use of His Majesty's ships.

Belfast Lough.

(c)   Harbour defences to remain in charge of British care and maintenance parties.

Lough Swilly.

(d)   Harbour defences to remain in charge of British care and maintenance parties.

AVIATION.

(e)   Facilities in the neighbourhood of the above ports for coastal defence by air.

OIL FUEL STORAGE.

Haulbowline (and) Rathmullen (:) To be offered for sale to commercial companies under guarantee that purchasers shall maintain a certain minimum stock for Admiralty purposes.

A Convention shall be made between the British Government and the Government of the Irish State to give effect to the following conditions:-

> That submarine cables shall not be landed or wireless stations for communication with places outside Ireland be established except by agreement with the British Government; that the existing cable landing rights and wireless concessions shall not be withdrawn except by agreement with the British Government; and that the British Government shall be entitled to land additional submarine cables or establish additional wireless stations for communication with places outside Ireland.

The lighthouses, buoys, beacons, and any navigational marks or navigational aids shall be maintained by the Government of the Irish Free State as at the date hereof and shall not be removed or added to except by agreement with the British Government.

That war signals stations shall be closed down and left in charge of care and maintenance parties, the Government of the Irish Free State being offered the option of taking them over and working them for commercial purposes subject to Admiralty inspection, and guaranteeing the upkeep of existing telegraphic communication therewith.

3.   A Convention shall be made between the same Governments for the regulation of Civil Communication by Air.

---

1 No. 214 NAI DE 2/304/1

Bibliography

MANUSCRIPT COLLECTIONS

British Library
Curzon MS
Walter Long Papers (WLP)
Public Record Office, Kew
Cabinet Records
House of Lords Record Office
Bonar Law Papers (BLP)
Lloyd George Papers (LGP)
University of Birmingham
Austen Chamberlain Papers
Bodleian Library, Oxford
Asquith Papers (Asq. P.)
H.A. L. Fisher MS
Lamar Worthington-Evans Papers (LWE)
Plunkett Foundation, Long Handborough
Horace Plunkett Diaries
Nuffield College, Oxford
J.A. Pease Diaries
Public Record Office of Northern Ireland
 Craigavon Papers
Spender Papers
Cabinet Papers
 National Archive of Ireland (Dublin)
Dail Cabinet Papers
Dail Debates
Trinity College Dublin
John Dillon Papers
Erskine Childers Papers
Richard Barton Papers
National Library of Ireland
John Redmond Papers
William O'Brien Papers
University College Dublin – Archives Department
Richard Mulcahy Papers
Eamon de Valera Papers (microfilm).

CONTEMPORARY NEWSPAPERS AND PERIODICALS

Belfast Newsletter
Daily Telegraph (London)
Freeman's Journal (Dublin)
Irish Independent (Dublin)
Irish News (Belfast)
Irish Statesman (Dublin)
Irish Times (Dublin)
Leader (Dublin)
Manchester Guardian
Morning Post (London)
Round Table (London)
Times (London)

PUBLISHED PRIMARY SOURCES

Hansard – Parliamentary Debates (5th Series) – House of Commons, House of Lords.
Addison, Christopher *Four and a Half Years* (London, 1934).
Amery, L.S. *The Case against Home Rule* (London, 1912).
Boyce, D.G. *The Crisis of British Unionism: Lord Selborne's Domestic Political Papers, 1885–1922* (London, 1987).
Butler, John 'Select Documents XLV: Lord Oranmore's Journal 1913–27' *Irish Historical Studies,* 29:116 (November 1995).
Chamberlain, Austen *Politics from Inside: An Epistolatory Chronicle, 1906–1914* (London, 1936).
Churchill, Winston S. *The World Crisis: The Aftermath* (London, 1929).
Correspondence between His Majesty's Government and the Prime Minister of Northern Ireland Relating to Proposals for an Irish Settlement (Cmd 1561).
Crowley, Jimmy *Uncorked!* (Free State Records CD 007, 1998).
Crozier, F.P. *Impressions and Recollections* (London, 1930).
Crozier, F.P. *Ireland for Ever* (London, 1932).
Dáil Éireann *Official Correspondence relating to the Peace Negotiations, June-September 1921*
Private sessions of Second Dáil
David, Edward (ed.) *Inside Asquith's Cabinet: From the Diaries of Charles Hobhouse* (London, 1977).
Esher, First Viscount *Journals and Letters of Reginald, Viscount Esher* (London, 1938).
Fanning, Ronan, Michael Kennedy, Dermot Keogh and Eunan O'Halpin (eds) *Documents on Irish Foreign Policy: Volume I, 1919–22* (Dublin, 1998).
George, William *My Brother and I* (London, 1958).
Good, Joe *Enchanted by Dreams: the Journal of a Revolutionary,* ed. M. Good (Dingle, 1996).
Gwynn, Stephen *John Redmond's Last Years* (London, 1919).

Gwynn, Stephen (ed.) *Letters and Friendships of Cecil Spring-Rice* (Boston, 1929).

Hopkinson, Michael (ed.) *The Last Days of Dublin Castle: The Diaries of Mark Sturgis* (Dublin, 1999).

Middlemass, Keith (ed.) Jones, Thomas *Whitehall Diary: vol. III – Ireland* (Oxford, 1971)

Morgan, Kenneth O. (ed.) *Lloyd George Family Letters* (London & Cardiff, 1973).

Northern Ireland Parliament Debates

Riddell, Viscount *Intimate Diary of the Peace Conference and After* (London, 1933).

Salvidge, Stanley *Salvidge of Liverpool* (London, 1934).

Taylor, A.J.P. (ed.) *Lloyd George: A Diary by Frances Stevenson* (London, 1971).

Welles, Warre B. *Irish Indiscretions* (Dublin, 1923).

Wilson, Keith (ed.) *The Rasp of War: The Letters of H.A. Gwynne to the Countess Bathurst* (London, 1988).

Wilson, Trevor (ed.) *The Political Diaries of C.P. Scott, 1911–1928* (London,1970)

Williamson, Philip (ed.) *The Modernisation of Conservative Politics: The Diaries and Letters of William Bridgeman 1904–1935* (London, 1988).

Young, Mary Alice *The Recollections of Mary Alice Young, née Macnaghton (1867–1946) of Dundarave and Galgorm, Co. Antrim* (Mid-Antrim Historical Group publication no. 32: Ballymena, 1996).

SUBSEQUENT PUBLICATIONS

Adams, R.J.Q. *Bonar Law* (London, 1999).

Bew, Paul *Ideology and the Irish Question: Ulster Unionism and Irish Nationalism, 1912–1916* (Oxford, 1994).

Blake, Robert *The Unknown Prime Minister: The Life and Times of Andrew Bonar Law, 1858–1923* (London, 1953).

Bowman, John *De Valera and the Ulster Question, 1917–1972* (Dublin, 1989).

Boyce, D.G. 'British Conservative Opinion, the Ulster Question and the Partition of Ireland, 1912–21' *Irish Historical Studies*, 17 (1970–71).

Boyce, D.G. *Englishmen and Irish Troubles: British Public Opinion and the Making of Irish Policy, 1918–22* (London, 1972).

Boyce, D.G., and Cameron Hazelhurst, 'The Unknown Chief Secretary: H.E. Duke and Ireland, 1916–1918' *Irish Historical Studies*, 20 (1977).

Budge, Ian, and Cornelius O'Leary, *Belfast: Approach to Crisis* (London, 1973).

Butler, David & Gareth *British Political Facts, 1900–1994* (London, 1994).

Callanan, Frank *T.M. Healy* (Cork, 1996).

Collins, Stephen *The Cosgrave Legacy* (Dublin, 1996).

Colvin, Ian, and Edward Marjoribanks, *Life of Lord Carson* (London, 1936) 3 vols.

Coogan, Oliver *Politics and War in Meath, 1913–23* (Maynooth, 1983).

Coogan, T.P. *Michael Collins* (London, 1991).

Coogan, T.P. *De Valera: Long Fellow, Long Shadow* (London, 1995).

Denman, Terence *A Lonely Grave: The Life and Death of William Redmond* (Dublin, 1995).

Dudley Edwards, R., & F.X. Martin (eds) *1916: The Easter Rising* (London, 1968).

Dutton, David *His Majesty's Loyal Opposition: The Unionist Party in Opposition, 1905–15* (Liverpool, 1992).

Eccleshall, Robert, and Graham Walker (eds) *Biographical Dictionary of British Prime Ministers* (London and New York, 1998).

Elliott, Sydney 'The Electoral Systen in Northern Ireland' (PhD, Queen's University Belfast 1971).

Emerson, Rupert *From Empire to Nation* (Cambridge, Mass., 1960).

Farrell, Brian 'Labour and the Irish Political Party System: A Suggested Approach to Analysis' in *Economic and Social Review*, 1 (1970).

Farrell, Brian *The Founding of Dáil Éireann* (Dublin, 1971).

Farry, Michael *Sligo, 1914–21: A Chronicle of Conflict* (Trim, 1992).

Fergusson, Sir James *The Curragh Incident* (London, 1964).

Fitzpatrick, David *Polititics and Irish Life, 1913–21: Provincial Experience of War and Revolution* (2nd edn., London, 1998).

Foster, R.F. *Modern Ireland, 1600–1972* (London, 1989).

Garvin, Tom *Nationalist Revolutionaries in Ireland, 1858–1928* (Oxford, 1987).

Garvin, Tom *1922: The Birth of Irish Democracy* (Dublin, 1996).

Gilbert, Bentley Brinkerhoff *David Lloyd George – A Political Life: The Organiser of Victory, 1912–16* (London, 1992).

Gilbert, Martin *Winston Churchill, Volume IV: 1916–22* (London, 1975).

Gwynn, Denis *John Redmond* (London, 1932).

Hart, Peter *The I.R.A. and its Enemies: Violence and Community in Cork, 1916–23* (Oxford, 1998).

Hopkinson, Michael *The Irish War of Independence* (Dublin, 2002).

Jackson, Alvin, *Home Rule* (London, 2003).

Jackson, Alvin *Sir Edward Carson* (Dundalk, 1993).

Jalland, Patricia *The Liberals and Ireland: The Ulster Question in British Politics to 1914* (Aldershot, 1993).

Jenkins, Roy *Asquith* (London, 1964).

Jennings, W.I. *Cabinet Government* (3rd edn., Cambridge, 1965).

Johnson, D.S. 'The Belfast Boycott, 1920–22' in J.M. Goldstrom & L.A. Clarkson (eds.) *Irish Population: Economy and Society* (Oxford, 1981).

Kendle, J. *Walter Long, Ireland and the Union, 1905–1920* (Glendale, 1992).

Kennedy, Liam, and David S. Johnson, 'The Union of Ireland and Great Britain, 1801–1921' in Boyce, D.G., & Alan O'Day (eds) *The Making of Modern Irish History: Revisionism and the Revisionist Controversy* (London and New York, 1996).

Laffan, Michael 'The Unification of Sinn Fein in 1917' *Irish Historical Studies* 17 (1971).

Laffan, Michael *The Resurrection of Ireland: The Sinn Fein Party, 1916–23* (Cambridge, 1999).

Lavin, Deborah *From Empire to International Comonwealth: A Biography of Lionel Curtis* (Oxford, 1995).

Lawlor, Sheila *Britain and Ireland, 1914–23* (Dublin, 1983).

Lawrence, R.J. *The Government of Northern Ireland: Public Finance and Public Services, 1921–64* (Oxford, 1965).

Lee, J.J. *Ireland 1912–85: Politics and Society* (Cambridge, 1989).

Longford, Earl of, and Thomas P. O'Neill, *Eamon de Valera* (London, 1970).

Lyons, F.S.L. *John Dillon* (London, 1968).

Lyons, F.S.L. *Ireland since the Famine* (London, 1968).

Macardle, Dorothy *The Irish Republic* (Dublin, 1937/New York 1965).

McCarthy, Andrew 'Michael Collins: Minister for Finance, 1919–22' in Doherty, Gabriel & Keogh, Dermot (eds.) *Michael Collins and the Making of the Irish State* (Cork, 1998).

McColgan, John *British Policy and the Irish Adminsitration* (London, 1983).

MacDowell, R.B. *The Irish Convention, 1917–18* (London, 1970).

MacLean, Iain *Rational Choice in British Politics: An Analysis of Rhetoric and Manipulation from Peel to Blair* (Oxford, 2001).

Mansergh, Nicholas *The Unresolved Question: The Anglo-Irish Settlement and its Undoing* (New Haven and London, 1991).

Maume, Patrick *D.P. Moran* (Dundalk, 1995).

Maume, Patrick *The Long Gestation: Irish Nationalist Political Life, 1891–1918* (Dublin, 1999).

Maye, Brian *Arthur Griffith* (Dublin, 1997).

Mitchell, Arthur *Labour in Irish Politics, 1890–1930* (Dublin, 1974).

Mitchell, Arthur *Revolutionary Government in Ireland: Dáil Éireann, 1919–22* (Dublin, 1995).

Mulcahy, Richard *Richard Mulcahy, 1886–1971: A Family Memoir* (Dublin, 1999).

Murphy, John A. *Ireland in the Twentieth Century* (Dublin, 1975).

Murray, Patrick *Oracles of God: The Roman Catholic Church and Irish Politics, 1922–37* (Dublin, 2000).

Nicholson, Harold *King George V* (London, 1952).

Ó Broin, Leon *Revolutionary Underground: the Story of the Irish Republican Brotherhood, 1858–1924* (Dublin, 1976).

Ó Broin, Leon *W.E. Wylie and the Irish Revolution, 1916–21* (Dublin, 1989).

Ó Corrain, Donnchadh (ed.) *James Hogan, Revolutionary Historian and Political Scientist* (Dublin, 2001).

O'Farrell, Patrick *Ireland's English Question* (New York, 1972).

O'Halpin, Eunan 'Historical Revision XX: H.E. Duke and the Irish Administration 1916–18' *Irish Historical Studies*, 29 (1980–1).

O'Halpin, Eunan *The Decline of the Union: British Government in Ireland 1892–1920* (Dublin, 1987).

O'Leary, Cornelius *Irish Elections, 1918–1977* (Dublin, 1979).

O'Leary, Cornelius *Celtic Nationalism* (Belfast, 1982).

O'Leary, Cornelius 'The Teaching of Politics in Ireland' in *Irish Political Studies,* 4 (1986).

Pakenham, Frank *Peace by Ordeal* (London, 1935).

Peatling, G.W. *British Opinion and Irish Self-Government, 1865–1925: From Unionism to Liberal Commonwealth* (Dublin, 2001).

Phoenix, Eamon *Northern Nationalism: Nationalist Politics, Partition and the Catholic Minority in Northern Ireland, 1890–1940* (Belfast, 1994).

Phillips, Alison *The Revolution in Ireland, 1916–23* (London, 1923).

Regan, John M. *The Irish Counter-Revolution, 1921–36: Treatyite Politics and Settlement in Independent Ireland* (Dublin, 1999).

Rowland, Thomas J. 'The American Catholic Press and the Easter Rising' *Catholic Historical Review*, 71:1 (1995).

Ryan, A.P. *Mutiny at the Curragh* (London, 1956).

Shannon, Catherine *Arthur J. Balfour and Ireland, 1874–1922* (Washington D.C., 1988).

Spender, J.A. and Asquith, Cyril *The Life of Lord Oxford and Asquith* (London, 1932) 2 vols.

Stewart, A.T.Q. *The Ulster Crisis* (London, 1967).

Townshend, Charles *The British Campaign in Ireland, 1919–21* (Oxford, 1978).

Turner, John *Lloyd George's Secretariat* (Oxford, 1980).

Wheeler-Bennett, J.W. *John Anderson, Viscount Waverley* (London, 1962).

Index

Sub-headings within individual entries are organized chronologically, not alphabetically.

The city of Derry and county of Londonderry are described by those names; the parliamentary constituency of Londonderry City is called by its official title.

Individuals are usually described in the text by the best-known form of their name, whether this is a proper name or a title of nobility; this index gives the unused name (or title) in brackets.

Although the Irish Unionists were part of the Conservative and Unionist Party (formed by merging the Conservative and Liberal Unionist parties in 1911) and the party as a whole was often described as 'Unionist', this book describes the British section of the Party as 'Conservatives' and the Irish as 'Unionists' for the sake of clarity.

'Nationalist' in the text refers to the followers of John Redmond and Joe Devlin; 'nationalist' refers to all varieties of Irish nationalist.